25-99

M-COMMERCE CRASH COURSE

P.J. LOUIS

McGraw-Hill Telecommunications

Ali	*Digital Switching Systems*
Ash	*Dynamic Routing in Telecommunications Networks*
Azzam/Ransom	*Broadband Access Technologies*
Azzam	*High Speed Cable Modems*
Bartlett	*Cable Communications*
Bates	*Broadband Telecommunications Handbook*
Bates	*Optical Switching and Networking Handbook*
Bayer	*Computer Telephony Demystified*
Bedell	*Wireless Crash Course*
Clayton	*McGraw-Hill Illustrated Telecom Dictionary* 3/e
Collins	*Carrier Class Voice Over IP*
Davis	*ATM for Public Networks*
Gallagher	*Mobile Telecommunications Networking with IS-41*
Harte	*Cellular and PCS: The Big Picture*
Harte	*CDMA IS-95*
Harte	*GMS Superphones*
Harte	*Delivering xDSL*
Heldman	*Competitive Telecommunications*
Macario	*Cellular Radio 2/e*
Muller	*Bluetooth Demystified*
Muller	*Desktop Encyclopedia of Telecommunications*
Muller	*Desktop Encyclopedia of Voice and Data Networking*
Muller	*Mobile Telecommunications Factbook*
Lachs	*Fiber Optics Communications*
Lee	*Mobile Cellular Telecommunications* 2/e
Lee	*Mobile Communications Engineering* 2/e
Lee	*Lee's Essentials of Wireless Communications*
Louis	*Telecommunications Internetworking*
Pattan	*Satellite-Based Cellular Communications*
Pecar	*Telecommunications Factbook* 2/e
Richharia	*Satellite Communications Systems* 2/e
Roddy	*Satellite Communications* 3/e
Rohde/Whitaker	*Communications Receivers* 3/e
Russell	*Signaling System #7* 3/e
Russell	*Telecommunications Protocols* 2/e
Russell	*Telecommunications Pocket Reference*
Shepard	*Telecommunications Convergence*
Shepard	*Optical Networking Demystified*
Simon	*Spread Spectrum Communications Handbook*
Smith	*Cellular System Design and Optimization*
Smith	*Practical Cellular and PCS Design*
Smith	*Wireless Telecom FAQs*
Smith	*LMDS*
Turin	*Digital Transmission Systems*
Winch	*Telecommunications Transmission Systems* 2/e

M-COMMERCE CRASH COURSE

THE TECHNOLOGY AND BUSINESS OF NEXT GENERATION INTERNET SERVICES

P.J. LOUIS

McGraw-Hill

New York • Chicago • San Francisco • Lisbon

London • Madrid • Mexico City • Milan • New Delhi

San Juan • Seoul • Singapore • Sydney • Toronto

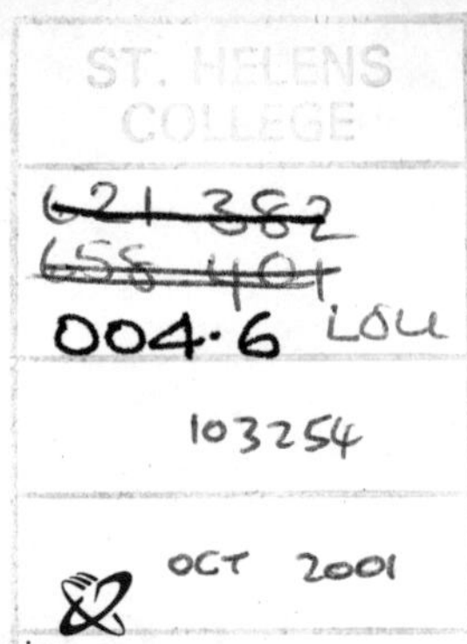

McGraw-Hill

A Division of The ***McGraw-Hill*** *Companies*

Copyright © 2001 by P.J. Louis. All rights reserved. Printed in the United States of America. Except as permitted under the United States Copyright Act of 1976, no part of this publication may be reproduced or distributed in any form or by any means, or stored in a data base or retrieval system, without the prior written permission of the publisher.

1 2 3 4 5 6 7 8 9 0 DOC/DOC 0 9 8 7 6 5 4 3 2 1

ISBN 0-07-136994-5

The sponsoring editor for this book was Steven Chapman, the editing supervisor was Steven Melvin, and the production supervisor was Pamela Pelton. It was set in Fairfield by MacAllister Publishing Services, LLC.

R. R. Donnelley & Sons Company was printer and binder.

McGraw-Hill books are available at special quantity discounts to use as premiums and sales promotions, or for use in corporate training programs. For more information, please write to the Director of Special Sales, McGraw-Hill, 2 Penn Plaza, New York, NY 10121-2298. Or contact your local bookstore.

Information contained in this book has been obtained by The McGraw-Hill Companies, Inc., ("McGraw-Hill") from sources believed to be reliable. However, neither McGraw-Hill nor its authors guarantee the accuracy or completeness of any information published herein and neither McGraw-Hill nor its authors shall be responsible for any errors, omissions, or damages arising out of use of this information. This work is published with the understanding that McGraw-Hill and its authors are supplying information, but are not attempting to render engineering or other professional services. If such services are required, the assistance of an appropriate professional should be sought.

This book is printed on recycled, acid-free paper containing a minimum of 50 percent recycled de-inked fiber.

DEDICATION

Dedicated to my darling wife Donna and our children Eric and Scott. Their love, tolerance, and support keep me going.

In memory of my late father Richard Louis. Who raised me in the best way he knew how; by giving me an appreciation for history, setting rules and goals for myself, learning from his mistakes, learning from others' mistakes, and learning from my own mistakes. He was unrelenting in his views. I miss him.

In memory of my late mother Jennie Chin Louis. Although she has been gone since 1966, she is still in my mind and heart. My late mother was an artist who painted, sketched, and sculpted. She passed on to me the ability to see/perceive the abstract as tangible.

In memory of the late Harry E. Young; my friend, superior, and mentor, who encouraged me to keep an open mind and be prepared for the unexpected. Harry opened a whole new part of the telecommunications business for me when he allowed me to team up with him.

DEDICATION

CONTENTS

CHAPTER 4: INTERNET— TECHNOLOGY AND SERVICE PROVISIONING IN THE INFORMATION MARKETPLACE 115

CHAPTER 5: NETWORK SIGNALING: MOBILE AND WIRED BUSINESS SPACE 153

ACKNOWLEDGMENTS

To Charles P. Eifinger, a 40 year veteran of the Bell System who retired in 1990, my friend and mentor; "Stop and think about what needs to be done. Then do it."

To the late Howard Schuster, Vice President of Engineering and Planning for New York Telephone (circa 1975–1985) and my former boss. He gave me the opportunity to be recognized.

To the late Carl Ripa, former chief engineer for New York Telephone (1985–1989) and my former boss. A fine leader who challenged me.

To Robert Mandell, my friend, superior, and mentor. His advice to me: "Don't sweat the small stuff," "Just Do It," and, "Cool heads prevail."

To Lawrence J. Chu, friend but more boss than colleague. To this day, he continues to challenge the boundaries of the telecommunications business; even after retiring from the Bell System.

To the *Telecommunications Industry Association* (TIA) and John Marinho, former Chairperson of TIA TR45 Committee, for teaching me the business of standards and expanding my view of wireless.

To all of my colleagues who have influenced me through our interactions and debates.

ACKNOWLEDGMENTS

Chapter 1

NETWORKS AND THE INTERNET: WHAT ARE THEY?

All telecommunications networks, regardless of the services they provide, share the following common characteristics.

1. Communications networks transport information (for example, voice, data, video) between the source and destination. Broadcast television and radio are examples where the information from one source is distributed to many different destinations.
2. Networks concentrate information. In other words, networks contain entities that concentrate information from multiple information sources prior to transmission.
3. Networks have entities that distribute information among multiple destinations.
4. Networks have entities that simply carry information from one point in the network to another. These elements are called transmission facilities or simply, facilities.
5. Networks may modify the information to improve transmission quality. Simple examples of this are analog-to-digital conversion, line coding, and modulation.
6. Networks exchange information with other networks. For example, when a caller initiates a long-distance call, the information flows through at least two local exchange

> networks and one long-distance network. Another example is when a wireless caller calls a wireline subscriber; the information flows through a wireless network and a wireline network. The Internet transports information between people and places. It supports the transmission (real time) of text, voice, and video. The traditional voice wireline network can support the same types of information as the Internet.

Local Area Networks (LANs), *Wide Area Networks* (WANs), *Virtual Private Networks* (VPNs), the Internet, and Broadcast television are just some examples of networks.

The terms "call" and "information" are used interchangeably. Whether information is voice, video, or an electronic text/graphic file, it is all information. Networks carry information from one place to another. The Internet is a medium for information exchange and is composed of highly intelligent terminal devices that communicate across low-speed and high-speed transmission paths. The Internet's power is in the terminal devices.

The wireless industry is an outstanding example of how the terminal device has become an integral part of the Internet business space. The devices currently in use appear to be the same mobile handsets that were used prior to the introduction of the mobile Internet. The user cannot physically see the differences. As far as the user is concerned, the handset looks and feels like the non-Internet models. The differences can only be seen via the services provided to the user. More information on the mobile Internet will be discussed later in this book.

Figure 1-1 is a diagram that provides a high-level view of a common telecommunications signaling network.

The reader will find that as the telecommunications network is generalized as an information network, the devices that enable information to be generated or transported are described in generic functions. Figure 1-1 describes the more traditional

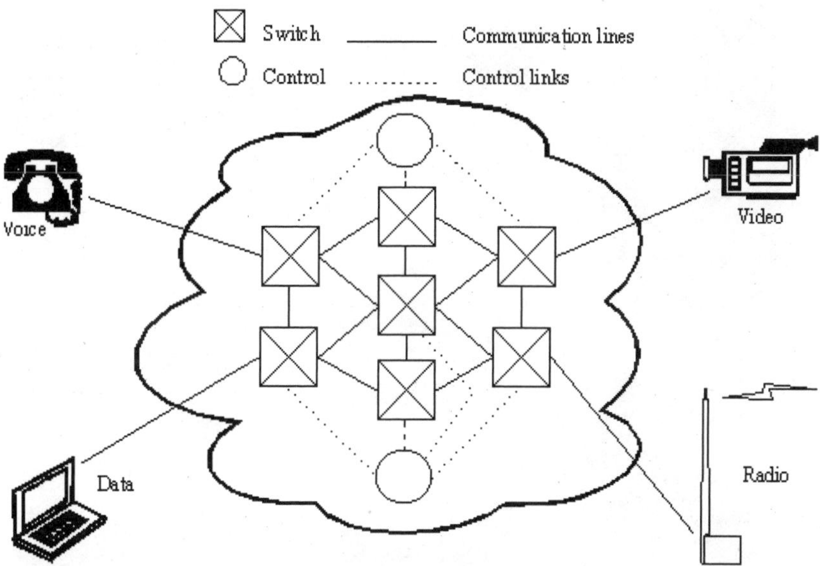

FIGURE 1-1 The telecommunications signaling network

telecommunications signaling network that carriers voice and data. This network diagram represents a non-packet and non-internet network. The Internet is similar in nature to current and more established wireline and wireless networks. The most glaring difference is that the Internet network is not controlled by a single large provider or even by a few large network providers. The Internet's power is in the terminal device. Figure 1-2 is a high-level illustration of the Internet and its basic components. It should be noted that the basic functional components of the Internet are similar to that of any other network.

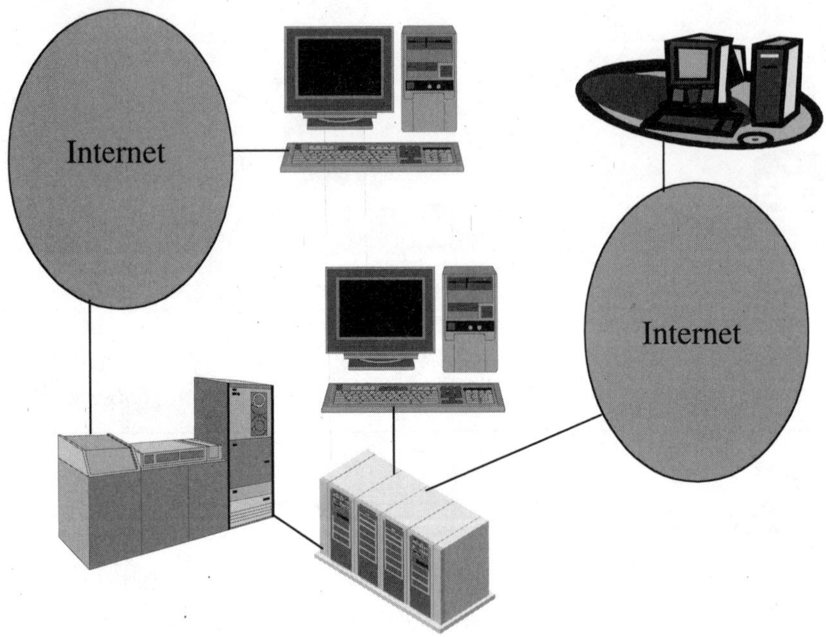

FIGURE 1-2 The Internet

NETWORK COMPONENTS

Telecommunications networks (including the Internet) are comprised of devices that tend to perform the same general functions. These devices or elements fall into the following general categories:

- Routers or switches
- Databases
- Transmission facilities
- Subscriber billing systems
- Customer care and provisioning
- *Network management systems* (NMS)

The aforementioned categories are applicable to both a wireless and wired (non-wireless) environment. These broad generalizations enable an individual to visualize and understand network architectures. As noted in *Telecommunications Internetworking*, by viewing the network as an entity that can be described as a set of functions rather than boxes with names, one will see the network as a dynamic entity. The Internet is the most dynamic, flexible, and user accessible network currently deployed today. The Internet does not require an *Internet Service Provider* (ISP) to have a spectrum license, file quarterly reports with a state regulatory commission, or maintain common service quality standards. However, this does not mean that the ISP is exempt from the aforementioned requirements. These types of requirements are driven by good or bad customer perceptions.

ROUTERS OR SWITCHES

Networks contain network elements that physically route information. These network elements are called switches or in the world of the Internet, they are called routers. Multiple devices may share the routing function. Depending on the type of network being addressed, the element may be a mainframe computer or a desktop computer. Routing is an activity or function. A router is a computer that directs information traffic among the network nodes from source to destination. Industry laymen often use their own industry specific terms in order to describe the same function. Routing or switching is sometimes separated by perceived network intelligence. In other words, routing implies intelligence and decision-making, whereas switching is merely a physical activity. This is a minor difference when one considers this from a high-level view.

Some networks employ fixed routing, meaning that a particular pair of users always communicates using the same circuit(s) or path(s) through the network. In other networks, dynamic routing is employed, and therefore, a different path is used each time. The Internet employs dynamic routing.

Unlike other types of telecommunications networks, the Internet has the bulk of its intelligence spread throughout the network in various terminal devices used by the customers.

Terminal devices can be as small as a laptop computer, which has so much processing power that the only limiting factor is the transmission network. Figure 1-3 illustrates the routing function.

DATABASE

Multiple routing elements may be required for a single connection. Routers use routing tables to determine how and where to send a particular information signal. These routing tables are communication roadmaps that reside in a database. The routing type employed in the Internet is dynamic routing. In other words, the Internet does not employ fixed or predetermined routing paths to route packets of data. The Internet database is a storage device for information. The database may be integrated within the host computer or exist externally to the computer, thereby serving multiple computing devices. When a single database serves multiple computers, the scenario is similar to that of the corporate Intranet. The database may even

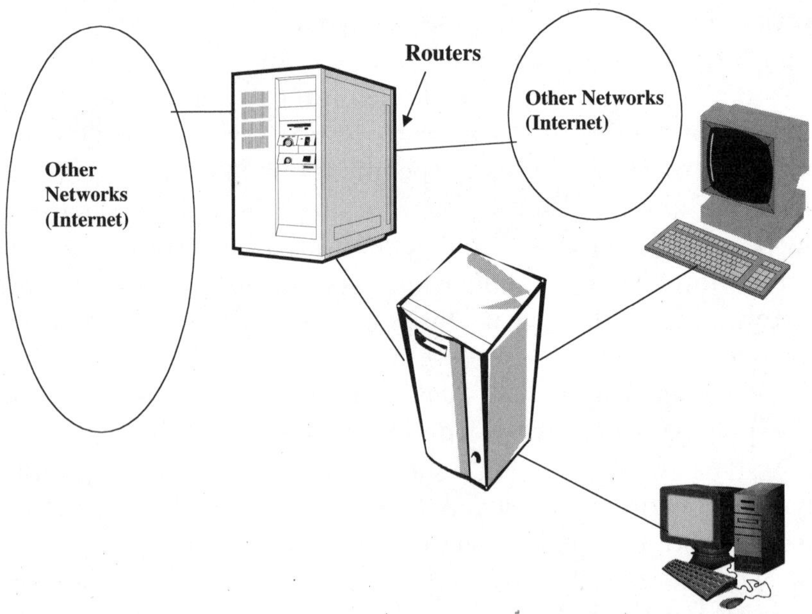

FIGURE 1-3 Routing and Internet component intelligence

store subscriber profile information, which may include billing preferences, subscriber features, restricted numbers, restricted Web sites, and so on. Figure 1-4 depicts the ways in which the database is utilized.

TRANSMISSION FACILITIES

As noted in *Telecommunications Internetworking*, transmission facilities are used to interconnect the end-user (subscriber) to the network, one network to other networks, and switching or concentrating transmission points within the network. Networks are often designed with a duplication of transmission facilities between two points. The Internet's ability to communicate with a multiplicity of Web sites is based partially on the way the transmission network has been designed. Today, high bandwidth networks are being constructed to support data. However, when the Internet began its commercial life, the network was a wireline telephone company designed and built

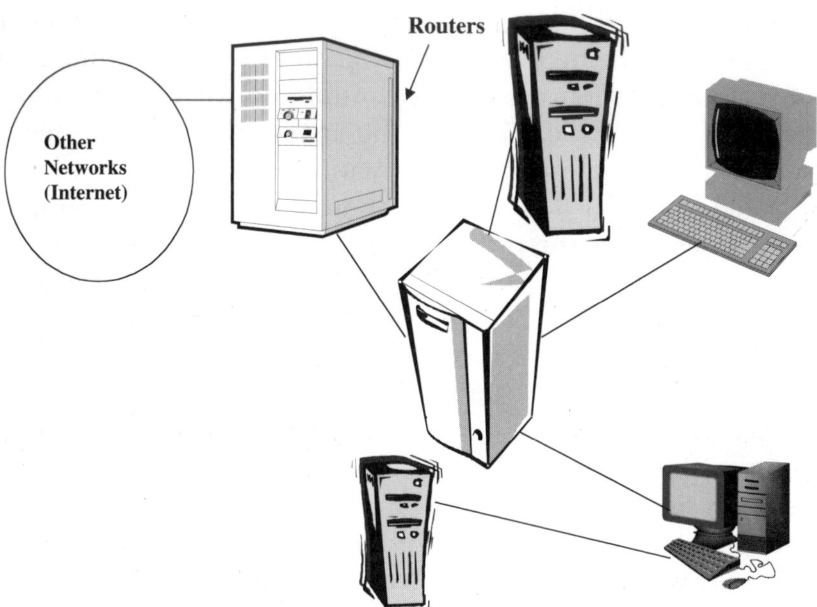

FIGURE 1-4 The database in the Internet

voice transmission network. The redundancy built into the older voice network enabled a large number of Web sites to communicate with one another with minimal difficulty. However, the Internet's dynamic routing strength was not being fully taken advantage of and was limited by the paths and bandwidth of the voice transmission network. The Internet is a packet data network. Packet data networks are dynamic routing networks. The wireless networks are employing hierarchical routing to route Internet traffic. The packet routing techniques have almost no value to the wireless network. More information on routing will be discussed later in this book.

The practice of multiple transmission path routing is called diversity routing and is used to improve reliability. Transmission facilities come in many different sizes and flavors–everything from 56 Kbps to many gigabytes per second. Transmission facilities are either metallic or glass. These facilities carry information in either analog or digital format.

The information carried by the transmission facilities is either out-of-band or in-band. An out-of-band information stream refers to a situation where call control information is physically carried over one path, while the actual call or content information is sent over another path. In-band information refers to a situation where call control and call/content information is physically carried over the same path in serial fashion. In the case of the Internet, information is carried in-band and digitized. Figure 1-5 is an illustration of the transmission network employed by the Internet.

NETWORK MANAGEMENT SYSTEMS

Service providers need some way of managing the various elements within their network. The management of the network elements may be done on-site or remotely. All systems must be managed. How that is done will vary from industry to industry and even within an industry. Network management practices and systems are as numerous as the number of service providers.

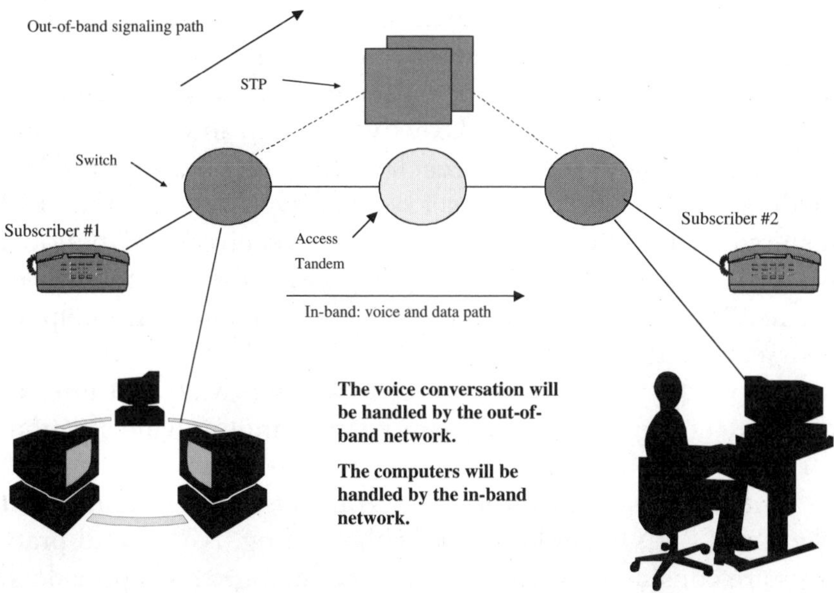

FIGURE 1-5 Transmission network

The majority of the ISPs currently maintain practices that are considered unheard of in the older wireline voice service provider or wireless service provider community. This refers to the frequently used ISP practice of taking the ISP network nodes off-line for maintenance. The older wireline voice service providers (the local exchange carriers) have a legacy of service that dates back to the nineteenth century, where service had to be maintained 24 hours a day, seven days a week. The wireless carriers, although not as old as the wireline carriers, have a philosophy of service that is the same as their wireline brethren. The mere thought that a customer could be removed deliberately from service on a periodic basis for maintenance was and still is considered "heresy." The reader should note that state regulatory agencies monitor the wireline carriers and not the ISP. Therefore, the standard of service for the ISP is lower. Remember that from the user's perspective and the regulatory agency's perspective, the standard of service is based primarily on availability, not bandwidth. Most user complaints about an

ISP are not about the bandwidth provided or the speed, but about the inability to gain access to the ISP. Not too long ago, in the late 1990s, a very large national e-mail ISP encountered the "busy signal" problem. Users would call into the ISP and discover that they could not reach the provider because the ISP modems were busy. The user's inability to access this ISP resulted in lost business, and the ISP received bad national press over the issue of inaccessibility (constant busy tones over the user's modem). A NMS would have been a massive help; of course, a NMS can only be employed if the ISP owns the physical network. Modems can report to a NMS where all lines to them are active; therefore, the modem cannot provide anymore service.

Industry standardization (*Transmission Management Network* (TMN) standards) notwithstanding, NMSs and practices are considered highly proprietary among service providers. An efficient NMS and set of practices may be the difference between a service provider's ability to show profit or none at all. NMSs enables a service provider to operate efficiently, control network element functions, and provide visibility to alarm conditions.

The *Competitive Local Exchange Carrier* (CLEC) is a type of carrier that was created as a result of the Telecommunications Act of 1996. The opening of the local loop has enabled new entrants into the business of providing local exchange services. One of the challenges that these CLECs have been facing is in the area of network management. An ISP is a type of CLEC: in other words, a data CLEC.

Starting up a CLEC forces one to focus on very specific areas of need first. These areas of need are identifying the market, business plan, marketing, core product, sales, and revenue. Unfortunately, network management was the last aspect of the business to focus on for many of the early ISPs. ISPs do not usually require a sophisticated NMS early in an ISP's operation; however, one should not expect to be in business for long unless a type of NMS is in place. Network management enables an

ISP to monitor a network and network element health and occupancy (level of activity). Figure 1-6 depicts the functions of network management.

SUBSCRIBER BILLING SYSTEMS

Service providers of all types need to be paid for the services rendered. All networks must record usage by the subscriber and bill the subscriber for the usage. Billing can be a function of usage volume, call volume, destination, or feature invocation. Service providers can offer service based on a variety of different parameters and subscriber activities. These parameters and subscriber activities may include calling party pays plans, toll-

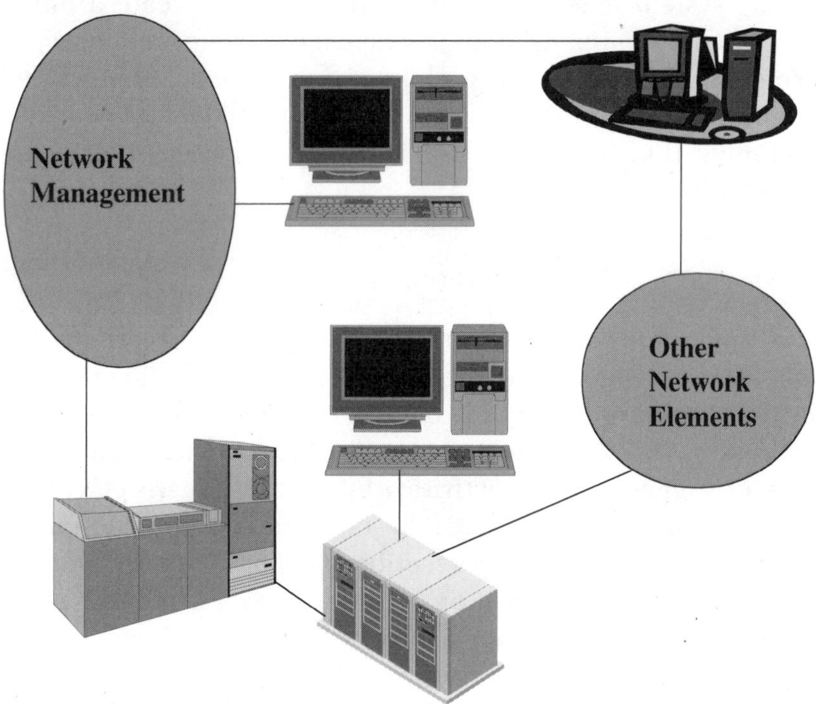

FIGURE 1-6 Network management

free numbers (that is, 800 and 888 numbers), credit card billing, geographic regional calling plans, time sensitive promotional plans, and other types of billing arrangements. The Internet is currently providing voice access through the use of *Voice over IP* (VoIP) technology. Most ISPs charge a flat rate for all of their services. The flat rate may only be used under specific calling conditions. The ISP changed the way telecommunications services are rated and billed. The paradigm for revenue generation in the world of the Internet is based partially on advertising and usage. This will be discussed in greater detail in later chapters.

As noted in *Telecommunications Internetworking*, billing is an area of operations that is often overlooked. Billing systems are treated as after thoughts. In fact, the billing system is as important as the switching/routing capability of the ISP. A billing system needs to do more than simply create a bill. It should apply company billing rates to a call, state, and local government surcharges, access charges, interconnect charges, and applicable discounts. These tasks apply to all service providers including the ISP. Figure 1-7 is a high-level depiction of the billing function.

Customer Care and Provisioning

Customer care and provisioning are activities and systems that may appear separate, but are directly related to billing systems. Both maintain a record of the subscribers' names and addresses, types of features subscribed to, and are used to support customer support activities by customer care personnel. These systems will be used by customer care personnel to answer questions about a subscriber's bill or the type of service, reflect payments, remove or add new services, request repair services, and most importantly, file customer complaints. Customers perceive all of this as a single system. However, they are typically separate, but interconnected data systems. Customer care is an activity that service providers of all types must support.

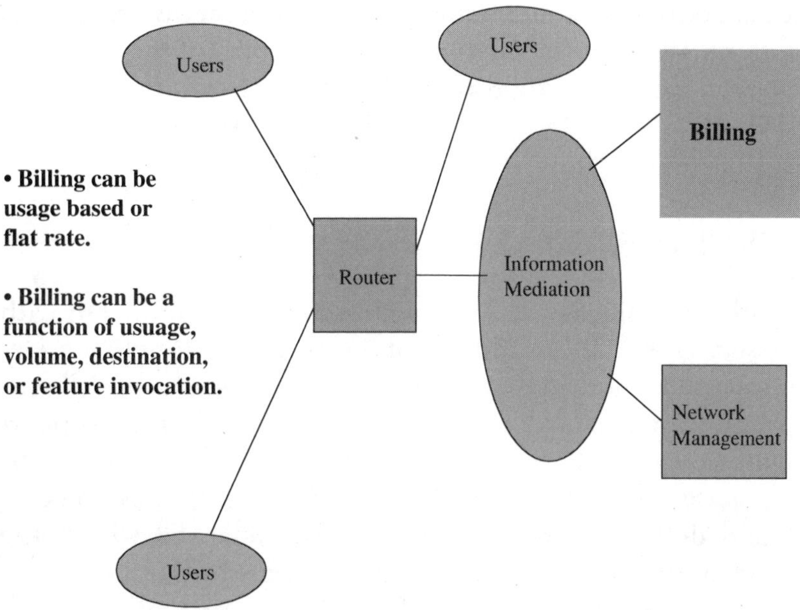

FIGURE 1-7 Billing

ISPs divide customer care into two categories. The categories listed are also supported by other types of carriers, but use other names.

Customer care/complaint The ISP interacts with the customer on the following types of matters: new service activation, billing inquiries, billing complaints, access telephone numbers, rating plans, and purchased feature capabilities.

Technical Support The ISP often maintains a separate "hotline" for technical questions. These technical questions can include ones concerning software installation, software troubles, access telephone numbers, and practically any question that the ISP cannot classify as a non-technical question. Often, the ISP will request that all questions be submitted to them via e-mail.

Successful ISPs understand that customer interaction is important. Without the customer, no revenue and therefore, no ISP exists. See Figure 1-8 for a depiction of customer care in the ISP.

NETWORK SIGNALING

Network signaling is the language used to communicate with other service providers and to communicate with network elements within one's network. As noted in *Telecommunications Internetworking*, two dominant forms of signaling to support network-to-network communications are available: *Multi-Frequency* (MF) and *Signaling System* 7 (SS7). In fact, the current and dominant form of communication used within the Internet is the same signaling used by the wireline telephone companies.

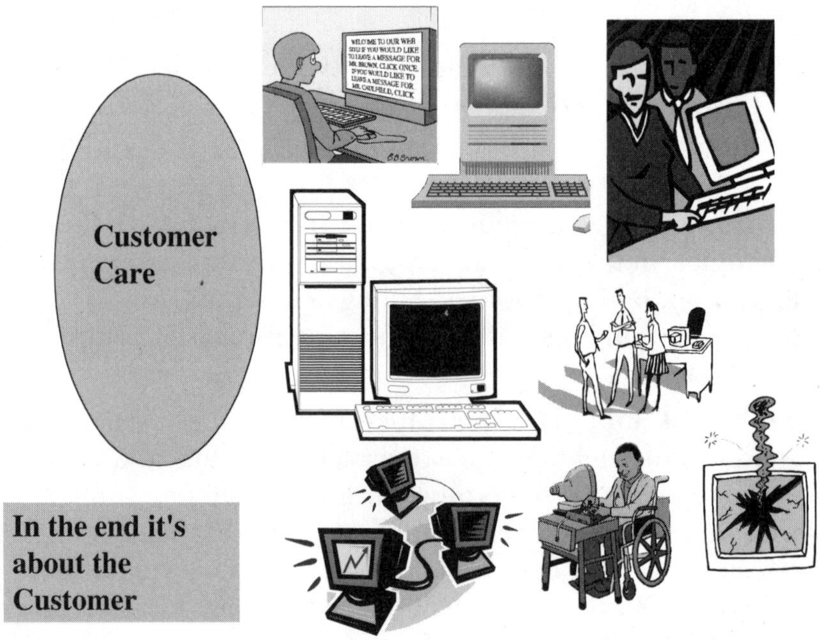
Customer Care
In the end it's about the Customer

FIGURE 1-8 Customer care and the ISP

Today, users have their computers dial a specific telephone number. The dominant and *incumbent local exchange carrier* (ILEC) allocates this telephone number to the ISP. Note that a national organization is responsible for the overall administration of telephone number exchanges (area codes) and the monitoring of local ILEC administration. Essentially, the ISP is treated like a customer asking for a telephone number. The ISP obtains numbers within various area codes to ensure that their customers do not have to pay long-distance access charges. Today, the ISP customer must pay the ILEC or CLEC for local access usage.

The user's computer communicates via a modem to the ISP's local host computer (server). The modem is a device that communicates via a series of tones. It is a modulator and demodulator device. The ISP can communicate with other ISPs via modems; however, it is more efficient for ISPs to communicate information between each other via a specific language. Note that until the late 1990s, the Internet could be characterized as a few large ISPs that provided only e-mail service and hundreds of small local/regional ISPs that also provided e-mail service. A specific ISP network signaling protocol is used called the *Transmission Control Protocol/Internet Protocol* (TCP/IP), which will be discussed in later chapters. Figure 1-9 illustrates how ISPs communicate with their users and with other ISPs.

THE INTERNET: WHAT IS IT?

The Internet was conceived in the 1960s. The United States government had retained the Rand Corporation to outline its vision of the future information network. The result was a report that described information as becoming a material good and the principal commodity of the twenty-first century. The U.S. government agency called the *Defense Advanced Research Project Agency* (DARPA) and invested several billion dollars in developing packet switching networks starting in the mid-1960s through the early 1970s. Initially, the customer for these

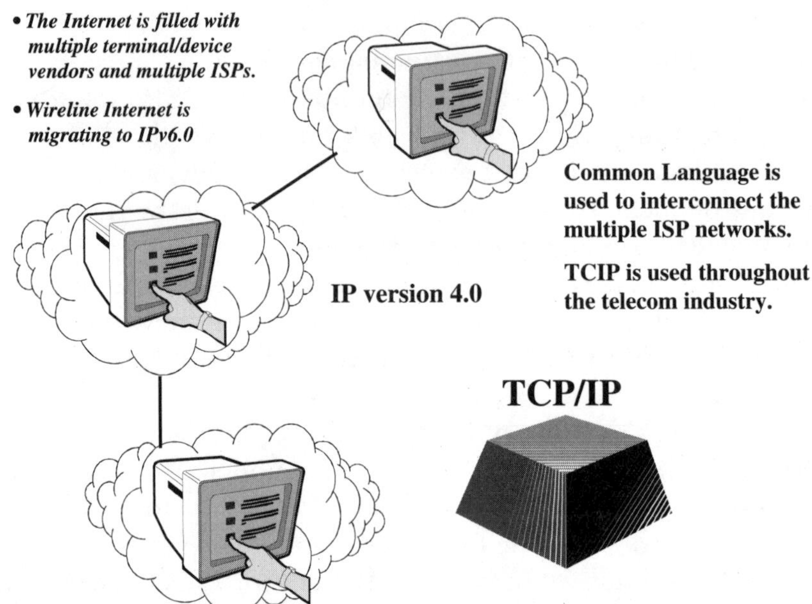

• The Internet is filled with multiple terminal/device vendors and multiple ISPs.
• Wireline Internet is migrating to IPv6.0
Common Language is used to interconnect the multiple ISP networks.
TCIP is used throughout the telecom industry.
IP version 4.0
TCP/IP

FIGURE 1-9 Network signaling

switching networks was the U.S. Department of Defense. However, many of the researchers who had been contracted to carry out the research and development were academics. These academics became enamored of the test systems they built and realized their work had applications beyond defense. As a defense initiative, this packet switching effort had a great deal to offer. During times of national emergency, relying on a public telecommunications network with known routing patterns/paths is not the most secure way of passing national security information. Initially, the research network was called the *Advanced Research Project Agency Network* (ARPANET), and later the Internet (Inter Network). The first e-mail was sent in 1972. Figure 1-10 is another rendering of the Internet.

By the early 1980s, two other important pieces of technology in the United States had emerged. The first was the workstation/server computer system, which emerged as the way to provide cost effective computing to the desktop. The second

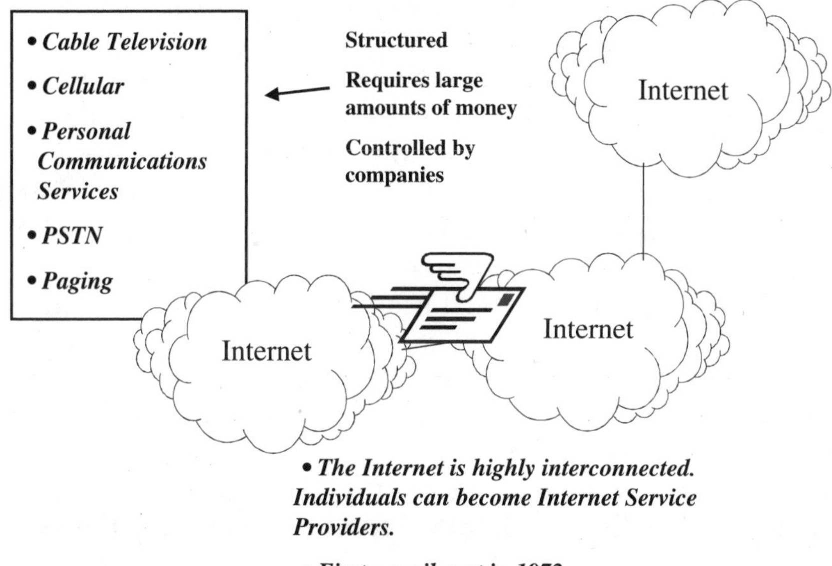

FIGURE 1-10 The Internet

was the Ethernet LAN. The Ethernet had been widely accepted as the way of providing communication between the desktop and server computers in the same organization (local area network). The computer and information technology researchers involved with Internet development had used all these systems for computing, local, and national communications. The United States government had funded most of the initial implementations of the Internet technology on the basis that it would be made freely available to others.

By the mid-1980s, a large number of universities and research labs in Europe and the United States had access to the Internet through largely government subsidized network links leased from the Public Network Operators. Today, most Internet connections are paid for directly by the subscribing organization. Unlike most telecommunications standards that are developed under the auspices of "recognized" groups (in other words, the *American National Standards Institute* (ANSI) and the *International Telecommunications Union* (ITU)), the Internet protocols were literally being developed out of sight. Even though the members of the ANSI and ITU are both large and small, massive corporate entities dominate ANSI and ITU in reality. During its initial development, the Internet was not a major economic force. Because no economic incentives existed for a large corporation to become involved in the Internet, ANSI or the ITU had no incentive to get involved.

The Internet was and still is an information network structure that is not controlled by a single entity. As a result of this lack of centralized control, most individuals in the telecommunications sector ignored the Internet. Internet protocol work is still defined outside of the jurisdiction of the dominant standards bodies. Current work on the *Wireless Access Protocol* (WAP) is actually done in other forums: specifically the WAP Forum.

Internet users and companies interested in providing Internet equipment recognized the need to develop protocol standards. Hence, the Internet Society, an International not-for-profit professional body, formed to give individuals, government agencies, and companies a direct say in the direction that the protocols and the technology could move. The Internet Society is not controlled by ANSI or ITU. Figure 1-11 illustrates the apparent free flowing nature of the Internet.

INTERNET NETWORK ARCHITECTURE

The Internet is an open pipe of information that allows access to anyone who uses a computer or some type of access terminal device. Currently, the Internet carries voice, data, and video (real time and stored). It has become the point of convergence for the telecommunications industry. Every telecommunications business segment is working diligently to find some way of obtaining the economic rewards of the Internet.

The Internet is not a single network, but rather a web of networks. One could also say it is a network of networks. Intelligence does not reside within a single component of the

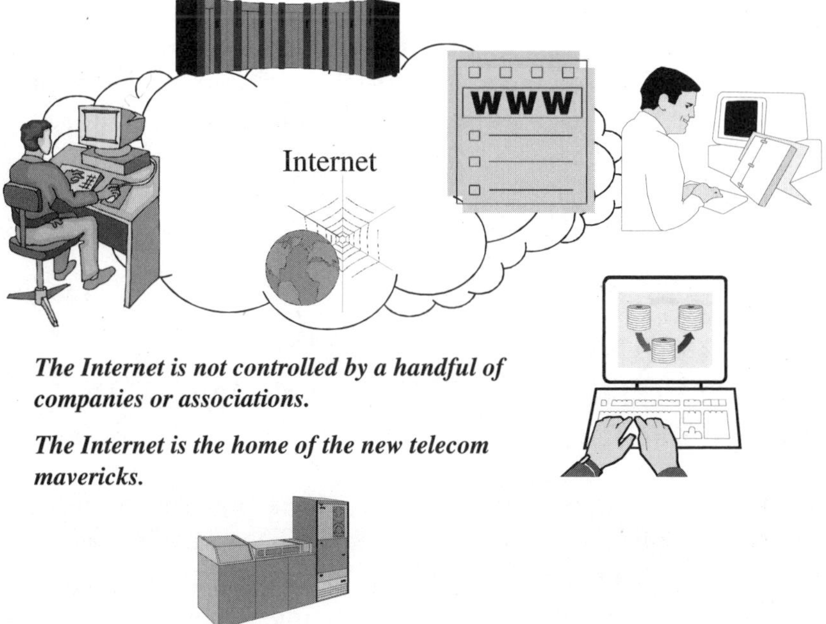

Internet
WWW
The Internet is not controlled by a handful of companies or associations.
The Internet is the home of the new telecom mavericks.

FIGURE 1-11 The Internet–Home of the new telecom mavericks

Internet, but is embedded within the components of individual network elements. When comparing the Internet with the older and traditional voice networks, one will find that the Internet is decentralized in its intelligence and control. The power of the Internet is in the software supporting the applications and within the protocol itself. The lack of centralized control has enabled many people who did not have the financial resources to start their own telecommunications business to create their own telecommunications business. The individual engineer has an opportunity to create a software application or build his/her own Web site with minimal financial investment or sacrifice. The small businessperson wanting to find a way of making some of the money that only large monolithic telecommunications carriers or e-mail companies made could now buy a low cost desktop and start his/her own local ISP serving the community in which he/she lived.

In the world of the Internet, no one controls the information pipe. This lack of control has created an enormous business opportunity for entrepreneurial engineers and business people. The Internet is quite literally unexplored territory for both the technical and business community.

Figure 1-12 highlights the major differences between the Internet and other telecommunications industry segments.

UNDERSTANDING THE NETWORK ARCHITECTURE: UNDERSTANDING THE MARKET FIRST

Another way of looking at the Internet is that the traditional public telecommunications networks (PSTN and Wireless voice networks) look at voice as their principal business and data (non-voice information) as a line of business they enter heavily. The Internet, on the other hand, looks at data (non-voice

- The Internet is not a service.
- The Internet is a network medium.
- The Internet is a medium that other telecommunications segments are attempting to utilize.
- The Internet was designed as a packet communications network with defense applications.
- The Internet's power is harnessed via software applications for terminal equipment.
- The Internet exists in both the wireline and wireless environments.

FIGURE 1-12 The Internet: An open pipe, opportunity, and unexplored territory

information) as its principal line of business and voice as a line of business it is beginning to enter. The traditional public telecommunications, carriers, satellite, and cable television are all seeking ways of entering the Internet business. Many of these companies do not even have a complete understanding of what they will be providing on the Internet. For example, a telephone company (wireline carrier) starts its own Internet business. The following questions need to be addressed within the wireline carrier:

- What is the wireline carrier providing on the Internet? Is it e-mail?
- How will the wireline carrier make any money? Will it charge for e-mail? Will it charge companies for advertising via banner advertisements? If the carrier seeks some of its revenue from advertising, is the Web site a kind of "electronic yellow pages?"
- Other existing e-mail companies are already charging low monthly flat rates. How will the wireline carrier bill for its services?
- What new services will the wireline carrier's engineers create that will enable the carrier to sell and charge money for?
- Who is the customer of the wireline carrier that is entering the Internet business? Does this customer already buy service from the wireline carrier? Does the wireline carrier need to conduct market research to find out who this customer is?

These may not seem like technical questions, but in reality, they are. Telecommunications engineers develop network capabilities and products for the sole purpose of making money for the company they work for. If the engineers are working without this kind of focus, the carrier will be "out of business." A brief side note: Product marketing, product management, and technology must work together in order to create services that people wish to purchase. The market consists of people who are seeking to have desires fulfilled. These people have their desires fulfilled by having their needs met through the creation, offering, and exchanging of products. Figure 1-13 illustrates the need for an Internet "wannabee" to understand the marketplace it is serving.

The challenge for the Internet "wannabee" is determining what product he/she will be providing. The Internet is not a product. It is an information medium that supports bi-directional transmission. Figure 1-14 illustrates these points. In terms of opportunity, the Internet is a "blank slate" waiting to be filled in.

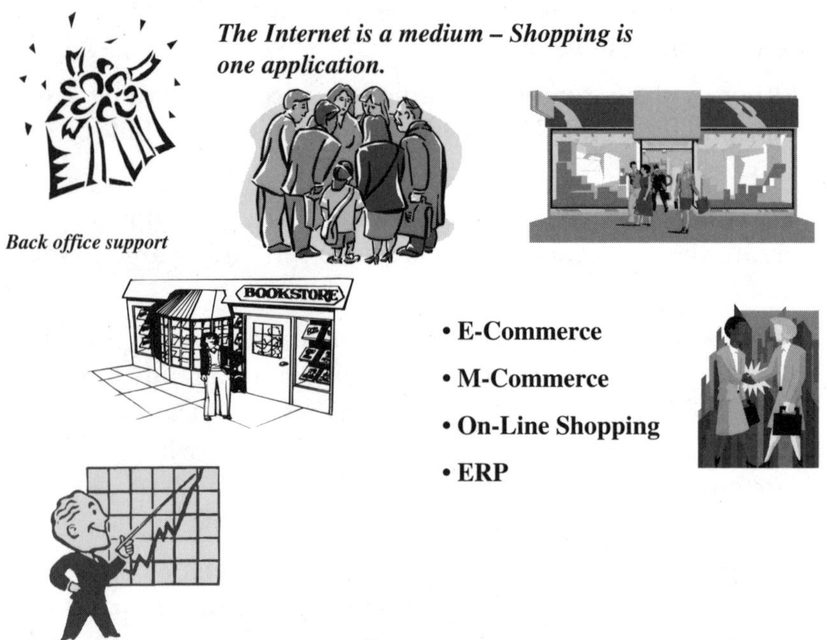

The Internet is a medium – Shopping is one application.
Back office support
BOOKSTORE
• E-Commerce
• M-Commerce
• On-Line Shopping
• ERP

FIGURE 1-13 Understand the marketplace: Creating services

The differences between the Internet business and other telecommunications business segments are deep; network architecture, traffic engineering profiles, and even the pricing mechanisms are different. This subject will be discussed in more detail later in the chapter. The hardware and software of the Internet can be defined within the functional framework described earlier in this chapter. The components and software used by ISPs is different than that used by the wireline and wireless (satellite, cellular, Paging, PCS, and so on) carrier communities. The following paragraphs outline the components used by an ISP.

• Web Browsing
• Research
• Publishing
• Commerce
• Entertainment
Physical newspapers replaced by OnLine news

FIGURE 1-14 The Internet–The new information medium

COMPONENTS OF THE INTERNET

The following sections describe the basic components of the Internet.

HARDWARE

The Internet is a collection of computing devices and transmission facilities communicating with one another. The Internet is basically made up of four pieces of hardware:

- Hosts
- Networks
- Routers
- Computers/Terminals with modems

HOSTS Hosts are computers that serve as central nodes for information processing and transactions. They can be workstations, PCs, servers, and mainframes that run applications. They are the central computer systems used to support users who wish to connect to a network or the Internet.

NETWORKS The networks are the transmission roadways built to connect a host and its subscribers to other hosts and their subscribers. In this case, networks are not any different conceptually than the traditional voice networks in a wireline or wireless voice service provider business.

LANs (Ethernets), point-to-point leased lines, and dial-up (telephone, ISDN, X25) links are all transport mechanisms that carry traffic between one computer and another. These glue together all the different network technologies to provide a ubiquitous service to deliver information packets.

ROUTERS Routers are special purpose computers that are good at talking to network links. Some people use general purpose computers as low performance (low cost) routers, that is, PCs or Unix boxes with multiple LAN cards, serial line cards, or modems. Routers are akin to wireline tandem switches. Routers direct traffic but are smart traffic switches that select the best route possible. They provide network management capabilities such as load balancing, dynamic route selection, trouble alert, and trouble identification.

COMPUTERS/TERMINALS WITH MODEMS Intelligence and capability within the Internet are embedded within the terminal devices; this includes the users' computers as well as the routers and hosts. The distributed intelligence and capability have enabled the Internet to flourish faster than the centrally controlled/owned networks. Think of this as analogous to moving a large mound of dirt 50 feet high and 50 feet wide from one place to another place by one person using only one shovel in a very short period of time. The amount of physical and mental energy needed to complete this task is enormous. (Completing this physical task in a short period of time requires similar concentration and focus as playing sports.)

Now imagine the same effort being accomplished by a thousand people each using a shovel. The overall amount of energy expended should be the same, but the difference is that each person will handle the shovel differently because each person will have a different level of experience in using a shovel. Some of the shovelers will have more experience using their body in physical activity and as a result, the effort will be accomplished far faster than one could calculate. These experience levels are random factors that a single large company could never tolerate in a large-scale development effort. Coordinated development efforts within a large corporate entity must work within a specific, defined process; this reduces waste and avoids disorganization. However, by tapping the total individual experiences of a population of engineers, the Internet business has exploded into the marketplace.

The result is faster computers, faster modems, unique network configurations, and unique applications of existing technology (for example, audio engineers who develop speakers for use in gaming technology thereby enhancing the market value of the computer). The power of the terminal exists because of the software developed. Software will be discussed next.

The typical network configuration of the Internet appears as follows. Figure 1-15 is a generic representation of the Internet. The Internet is a conglomeration of computers and LANs. All of the above is designed to communicate with the other.

SOFTWARE

For many people in the non-technical community, software refers to games, word processing programs, or graphics art programs. For many in the technical community, software refers to games, word processing programs, engineering design programs, or CAD/CAM programs. These programs are the products that helped generate interest, revenue, and investment

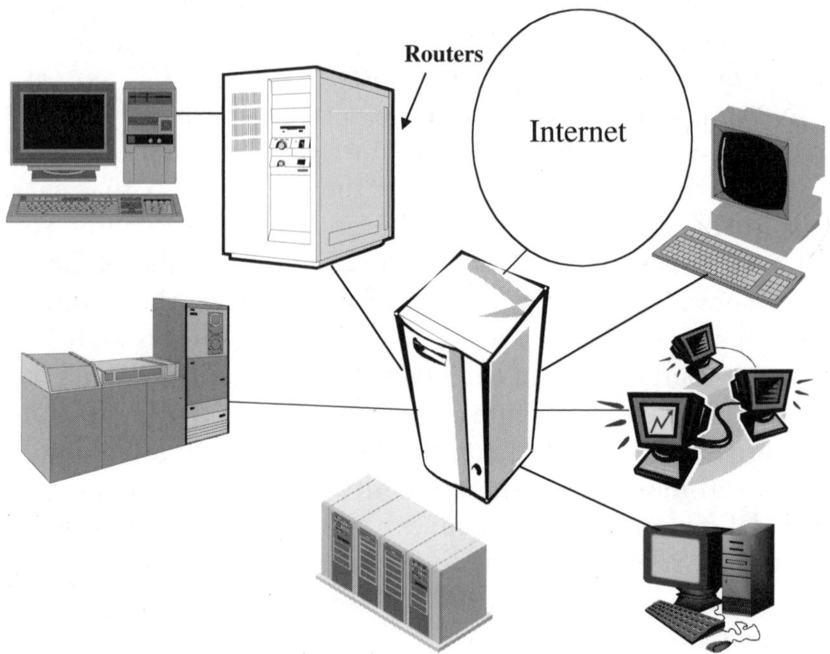

FIGURE 1-15 Typical network configuration

in the Internet business. However, the underlying software that has facilitated the growth of the Internet is the software that enabled banner advertising and interactive icons.

Banner advertising jump-started the revenue stream for the ISP. The ability for a user to interact with an icon in order to move from one screen to another or to initiate an activity made the Internet a user friendly place for the non-technical and technical (but not computer literate) to work and play. Software is no different than any other product. Different types of software are available such as user software (like a word processing program or a utility program), operating software, game software, software that runs a telephone switch, software that creates a voice service like Call Forwarding, and software that enables a Web site to interact with a user. Figure 1-16 depicts the various types of software from a functional perspective. Later chapters will discuss software in more detail.

Software is needed to operate and run the system.

Software is needed to facilitate the use of the equipment.

- **Operating System Software**
- **User Applications**
- **Utility Software**
- **Web Scripting**

FIGURE 1-16 Software

INTERNET ACCESS AND NETWORK INTERCONNECTION

The Internet is accessed and used by corporate users and home users. The corporate user typically accesses the Internet via their corporate Intranet. The corporate Intranet interconnects the company LAN via a host. Further information on the Intranet is available later in this book. The home user accesses the Internet via an ISP. The host uses TCP/IP to access the Internet. Today, the home user accesses via a modem in the home computer.

As indicated earlier, the Internet is a network of networks. Each network communicates with the other using computers as the principal terminal device. The corporate user is using a data network to communicate with the Internet also known as the World Wide Web or the Web. Typically, the data network is connected to the PSTN. The home user purchases access from an ISP. The ISP serves as the host to subscribers of their service. The ISP is reached by home computers using modems that dial telephone numbers associated with the ISP. The modem makes these ISP calls on twisted pair wire. Even though the Internet is a packet-switched network, the link between a home PC and an ISP through a modem is still circuit-switched. New technology, such as DSL, is changing this environment. DSL makes it possible to have Internet access "always on." The ISP and the corporate Intranet are connected to several Domain Name Servers (DNSs). The DNSs enable the users to find the destination/requested Web site (another host). Figure 1-17 illustrates the standard network interconnect.

The aforementioned represents two distinct ways to connect the various Web sites, yet a common node still exists called the *Packet Switched Telephone Network* (PSTN). The role of the PSTN is changing in the world of Internet. The wireless carrier community is working to enable the mobile handset to interact with the Internet. The PSTN plays a major interconnect role to date. However, the Internet is not a single entity or a manageable number of entities that can be controlled by the PSTN (this includes CLECs, ILECs, or interexchange carriers). The

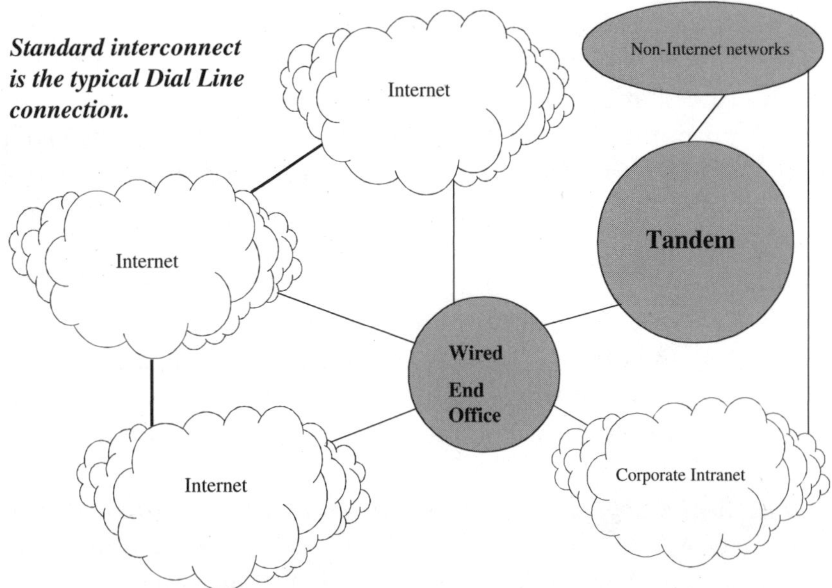

FIGURE 1-17 The typical network interconnect

construction of a network to support Internet users can take the form of a single ISP owning telecommunications hubs throughout the nation while leasing the nationwide transport from another company (that may not even be a major long distance company or an ILEC).

The wireless Internet is evolving differently than the wireline-based Internet. Wireless license holders are converting their networks to support either the WAP or i-mode signaling protocols. The proliferation of wireless devices has caused those in the wireline worlds to worry about lost market share. Rather than being the second communication device in the home, the wireless device is being positioned to take the lead as a user's principal way of communicating. The wireless Internet has not only expanded the wireless business, but it has also impacted the network interconnect to the wireline community. Network interconnect to the wired community becomes less essential as the wireless base not only grows, but also becomes a more important form of communication. Even more so, no industry work has been done to address network interconnect to the wireless Internet. The wireless Internet is evolving as a standalone network. Given its growth, the need for interoperability (interconnect) may be unnecessary.

The corporate Intranet is a LAN that is an effective internal company Internet. The Intranet supports everything that the Internet user has access to, but its principal function is to support corporate communications. The Intranet obtains a dataline, probably a facility at T1 or T3 rates, from the wireline telephone company. The Intranet is most likely interconnected to an ILEC end office. The wireline telephone company assigns a dialable number to the company. The dialable number is associated with a domain name and address that hosts can access via their TCP/IP connection.

The home user accesses the ISP via dialable numbers. The home user essentially calls the ISP via the home computer and reaches other Web sites through this ISP. The standard interconnect to the Internet for an ISP is via a typical dial line from the end office. Today, the telephone companies have installed routers in their networks that do nothing but interconnect Intranets and ISPs. Network interconnect in the realm of the Internet is no different than that of any other type of carrier. The potential business spin on ISP network interconnect is that all of the other telecommunications segments are working towards the Internet model. The convergence toward the Internet model will force all other carrier types to expend resources for developing a Web presence of some kind. Figure 1-18 is an illustration of a generic network interconnect involving the ISP.

See Appendix B for information on network interconnect. At this time, the only industry standards and defacto standards that dictate network interconnect are Bellcore's (now Telcordia) GR-145-CORE specficiation and the *Telecommunications Industry Association's* (TIA's) ANSI standard ANSI-93. It should be noted that network interconnect is about communication between networks. Network interconnect does not need to address user technical interaction with an ISP. Users always interact directly with their service provider. The issue is how the user's ISP communicates to other networks; this is network interconnect.

ISPs are interested in network interconnection as much as any other carrier. They are required to pay for long-distance access. Every time an ISP reaches out into the Internet web outside of the LATA boundaries of the ILEC, the ISP is required to pay an access charge to the ILEC. Some ISPs usually pass on the cost of the long-distance portion via some type of long-distance transport charge. This charge would be over and above any hourly fee the ISP would levy on the subscriber/user for use of the ISP services. Note that some ISPs simply charge the subscriber/user for access to the ISP services and hold the subscriber/user responsible for the cost of the local telephone call via the normal monthly telephone bill.

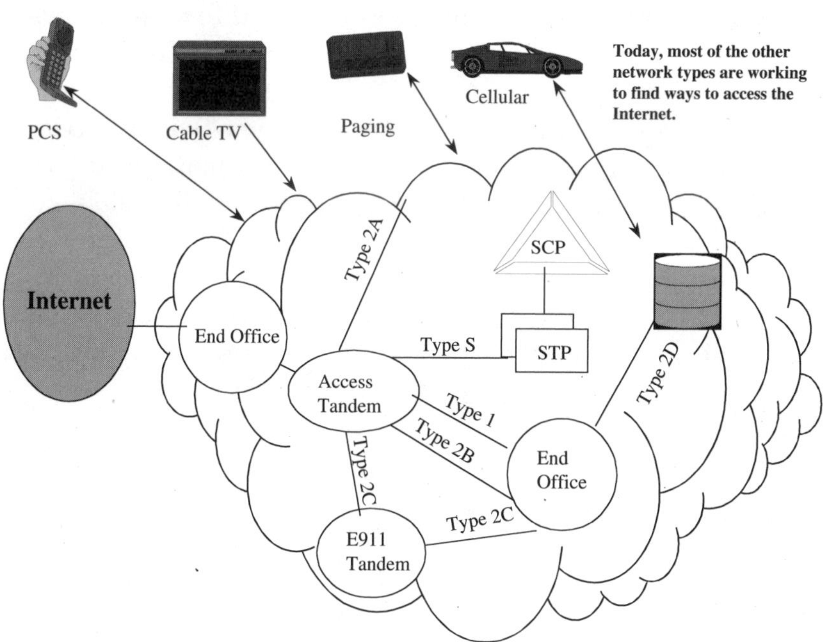

FIGURE 1-18 The Internet and network interconnect

Anyone can be an ISP. Many understand this and as a result, the Internet has been flooded with a plethora of ISPs. ISPs can serve local communities, regions of the country, the nation, and provide international access. The word "web" was used to describe the Internet. The thousands of ISPs and Intranets, if interconnected to one another, would form a picture of a spider web. Many people speak of the Internet backbone. This is a hard scenario to imagine, for no single company is in control of the Internet. However, long-distance carriers who own nationwide transmission networks and non-telecommunications companies who have internal communication networks capable of supporting telecommunications traffic of other companies exist. In this way, the Internet does have a backbone. Figure 1-19 is an illustration of the "loose" nature of the Internet business. This situation will change as content monitoring becomes a necessity in order to protect children from various unsavory influences.

Without connectivity to other networks, the Internet rapidly loses value. Figure 1-20 emphasizes the need for a nationwide backbone network in order to support nationwide Internet connectivity.

The Internet today is marketed as an enabler of services, better services, point of sale, e-mail, and information research. Remember that the Internet is accessible by all and is controlled by no one. Even though ISPs must seek ILEC interconnection, the fact remains that the growing CLEC community is slowly taking away the ISP interconnect business. Therefore, network visibility is and will be maintained through the CLECs. Figure 1-21 illustrates the role of the CLECs.

INTERNET MODELS

A variety of business models and network configurations are supported in the Internet community. The great challenge for Internet players is determining what it is that they will provide. Because no one company owns or controls the Internet and the job of the carriers is to sell voice and data usage on their net-

FIGURE 1-19 The Internet is a web of various companies

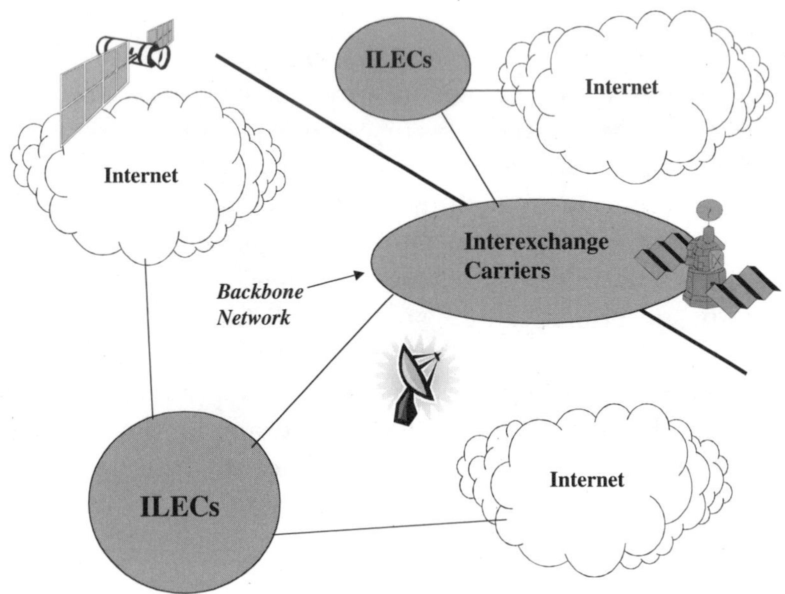

FIGURE 1-20 Nationwide connectivity

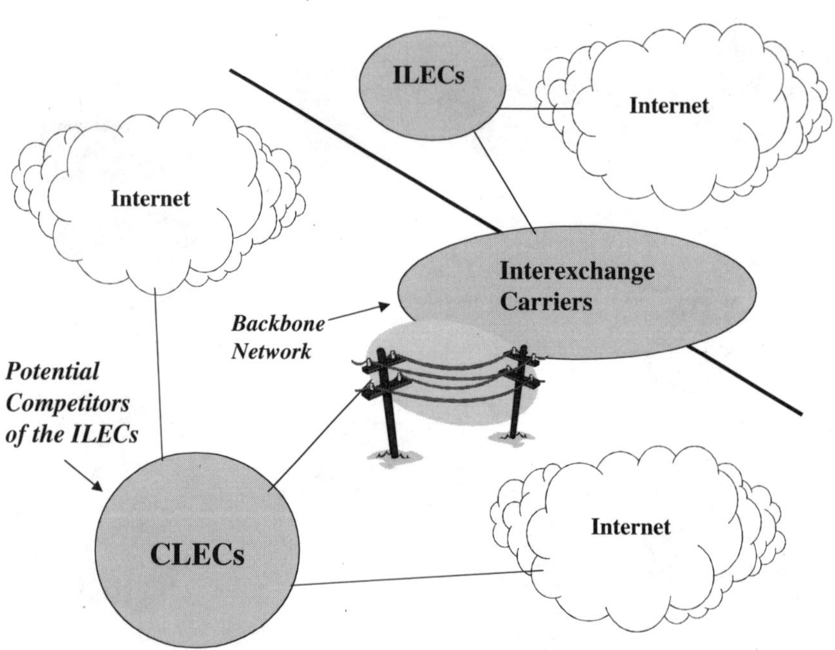

FIGURE 1-21 CLEC's and the Internet

works, what exactly are the carriers providing with their Web sites? Are the carriers providing physical products, access to purchasing physical products, or other types of services? The following chapters will look to answer these questions further.

The Internet is a medium that enables communication between users of computers. The Internet is not an application per se, rather the Internet is a medium that is a blank slate as it relates to subscriber value. No one individual can say what product the Internet provides. To many, it seems to be the principal form of communication. The Internet supports a variety of capabilities, but it is not a product. For example, the following are functions, not products and services, of the Internet. These functions are not purchased by subscribers:

Research Library access.

Advertising Advertising space is sold as banners on a Web page. No ISP subscriber purchases the service of an ISP simply because it has "cool looking" interactive point and click banners. The advertising might attract them, but it certainly will not generate "solid, recurring, and dependable" revenue for the ISP.

Web page publishing This refers to any type of text material that can be found on a Web page.

Applets Software applications that are either found on the computer or downloaded off of a Web site for use at that moment.

Web Browsing Search engines are used to search in an organized fashion for other Web sites or topics. URL support.

Communication between people and computers E-mail.

Figure 1-22 highlights the various applications of the Internet.

Even though the aforementioned services are not purchased by Internet users, they are still value added capabilities that make the Internet attractive to users. The question still remains, what products and services do the users buy from the Internet? The answer is that the Internet itself is not a product, but it provides a service, which is access to people and information (databases). This access is what makes the Internet a valuable tool. The information obtained may be used for a variety of purposes including purchasing products from the database holder. The user is limited by what the targeted database will allow and by the software on either the desktop computer or the targeted host site.

The purchasing of products is an activity that has come to be known as *e-commerce* (Electronic Commerce). E-commerce

Functions not product

- Research
- Advertising
- Web page publishing
- Applets
- Browsing
- E-mail

FIGURE 1-22 Applications

is catalog purchasing off of the Internet where the selling party displays pictures of its products, interactive (point and click) icons, and provides online purchase order forms. These product sellers even have their databases tied into credit card validation databases. E-commerce is an activity/application and not a product to be sold to an Internet user. E-commerce has grown to become the buzzword describing all of the Internet money making business activity. It has become the model for many companies wishing to use the Internet for doing business and to base their business models on. The challenge for the ISP is determining how it makes its money from e-commerce. Does the ISP charge the retailer a fee per e-commerce Web site hit?

Does the ISP charge the retailer a flat fee? How does the ISP make money from e-commerce? E-commerce has generated tremendous interest and revenue from the marketplace, but it has caused most of that revenue to flow into the hands of the retailer. At the time of this writing, m-commerce is fairly new to Internet business space. M-Commerce can be described as mobile e-commerce. The impact of the mobility attribute has yet to be fully understood.

Figure 1-23 highlights how the Internet has become a medium for retail shopping and making money for the ISP.

The question remains, what business models can the ISP base its business on? The following chapters will examine the various models that the ISP community can follow.

FIGURE 1-23 Online shopping and e-commerce

SUMMARY

The Internet can be characterized as a wide-open playing field. The large established mega-carriers do not control the Internet. The mega-carriers may own a lot of network infrastructure, but they still are working to determine how to make money off of the Internet. The ISP faces many challenges that are no different than any other company trying to provide some kind of service to the marketplace. The ISP has the following concerns:

- Identifying the market segment to serve.
- Identifying the product/service to sell.
- Creating differentiation between itself and other ISPs.
- Identifying the location to launch the service.
- In general, determining what business the ISP is in.

The Internet has become a massive opportunity for software developers, retailers, computer manufacturing companies, computer retailers, banner advertising companies, and venture capitalists. Early in its birth, the Internet was making money for everyone involved, even if the Internet company was providing another e-mail service. However, as the activity on Wall Street has shown, the patina on the Internet business has become a little tarnished. Investors were seeking revenue originally and not necessarily profit. The financial community has done a "mental reset/reboot" of its perception of the Internet. Investors are seeking longer-term plays and commitments from the entrepreneur, which has resulted in hard "looks" at the business models of the various ISPs and other Internet companies. However, one should not worry about the "tarnish" on the Internet business because the Internet is here to stay. Intuitively, one can see the opportunity the Internet brings–access to the people and information access by the people. The challenge for the ISP is to figure out how to make a business out of the Internet. The ISP is the new carrier of the future.

The following chapters will look at the entire Internet business and will examine the technical and business pieces of the ISP.

CHAPTER 2

BUSINESS MODELS

The years 1999 and 2000 have been marked with financial change for the entire CLEC community including the ISPs. Whereas once the ISP was able to obtain money from *venture capitalists* (VCs) for simply saying they would provide e-mail, the money became scarcer in 1999. In 2000, the money became impossible to obtain unless that ISP business plan had substance.

Today, a startup Internet company, no matter what it might be doing or providing for the Internet, cannot compete unless some substance backs the business plan. Until 1999, a startup ISP could simply advertise that it was supplying e-mail service with great looking graphics and a free prize just for signing up. The startup ISP could execute its business plan by such methods because the marketplace was still in a learning phase. The customer was willing to try every flavor and kind of ISP service. The customer had become more daring because the dominant e-mail service company (no names shall be given) had inflicted the worse kind of trouble on it subscribers–the busy tone. During 1997-1999, the dominant e-mail service company was unable to provide access for most of its subscribers unless the customer dialed into to its server at 2:00 A.M. The lack of access drove millions of subscribers to other e-mail service companies. To this e-mail company's credit, its single rate plan (with unlimited usage) exponentially grew its customer base. This lack of access created a new opportunity for the Internet "wannabees."

The new startup *Internet Service Providers* (ISPs) took advantage of the situation and launched their own e-mail services. These services were local, regional, and national in flavor. Until 1999, these new ISPs were primarily either local or regional. However, these startups presented enough of a challenge to the dominant national ISPs that the venture capital community poured billions of dollars into the Internet market. The financial community's outpouring of interest helped spur massive growth in Internet technology and business. The massive mistake of one e-mail company did not serve as the primary driver for the growth of the Internet, but it did open up an opportunity for new money to find its way into Wall Street. Figure 2-1 is a rendering of the financial community's interest in the Internet.

FIGURE 2-1 The growth of the Internet

THE RELATIONSHIP BETWEEN TECHNOLOGY AND BUSINESS

Growth in any technology or business is a relationship. Technology that needs money for development must have a business reason for the investment. A business dependent on new technology requires financial investment in the technology.

Investors are now asking ISPs and other Internet-related companies, "What is your company providing over the Internet or to the Internet customers?"

It may seem a little odd to ask such a question, however, until 1999, Wall Street was dumping money into any company that had the words "dot com" in its name. Many companies in the xDSL provisioning business even benefited because the investment community believed that they were in the "dot com" business. To the casual observer looking in from the outside, the "dot com" was all the rage and no one was actually asking the kinds of questions an investor should ask of any new opportunity. The kinds of questions that should have been asked of these new startups and their founders were:

- What are you selling?
- Why would anyone want to buy it?
- What makes you think they will buy from you?
- How big is the market?
- Are you already operating as a company?
- Are you looking for seed capital or are you further along in the search for financing?
- How much of this market do you expect to capture for your company?
- Why do you think you will be successful?
- What do you plan on doing with this company? Do you plan on taking it public?

- If you plan on going public, when do you expect to do so?
- If you do not plan on taking the company public, what do you plan on doing in increasing value to your investor?
- What does your team look like? (In other words, who are your team members?)
- What do you want from me (the investor)?
- What are you bringing to the bargaining table other than a good idea? Do you have any "skin" in the game? "Skin" refers to an individual's personal stake in the business (usually the stake is one's life savings).
- How much money do you want?
- How much of the company will I get if I invest in you?

In the year 2000, the investor has become very "Internet savvy" by asking all of these questions, which will each have a sub-listing of questions. The people seeking financing must be able to answer all of the questions. However, it is important to note that the "idea" being promoted must get the attention of the investor. If the "idea" does not capture the interest of the investment community, the entrepreneur wannabee will not even hear the rest of the questions that need to be asked. Many of the pre-2000 startups did very little in terms of returning any value to the investor. In fact, many startups simply lined the pockets of the founders with money. Figure 2-2 depicts the investor's relationship with the startup.

BUSINESS MODELS

The Internet business models to date can be classified in two broad categories:

- ISP
- Access

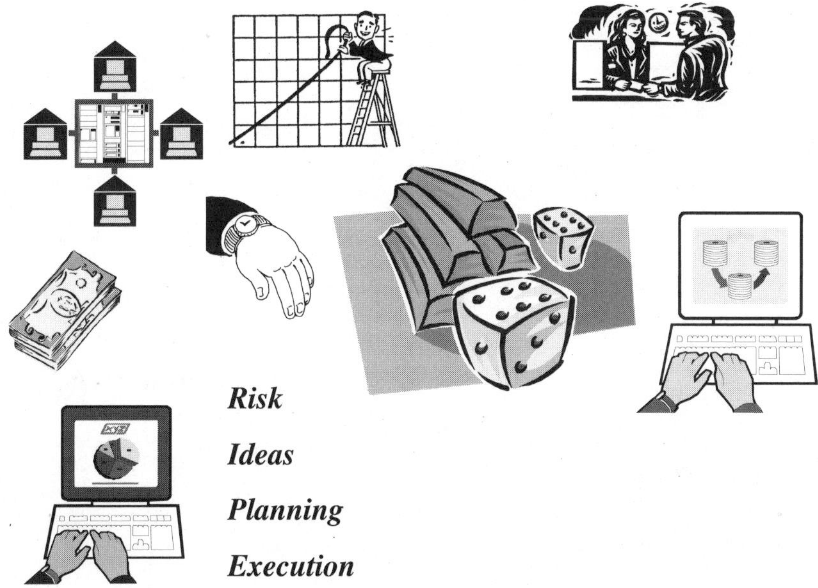

FIGURE 2-2 The investor and the startup

Figure 2-3 depicts these two categories of business models. Within the two categories of models are sub-categories of models. This section of the book will lay the groundwork for Chapter 3, "Challenges of the Internet Business-Business and Technology."

Internet Service Provider (ISP)

The ISP is generally the company that provides the service directly to the user/subscriber. The ISP can be even further broken down into sub-categories of ISP types. This is similar to the traditional telecomm service provider community. Several different kinds of service providers are available; some provide local service or long-distance, some provide both, and so on. In the book *Telecommunications Internetworking*, the wireline telephone carriers were referred to as the traditional telecom player. In this book, wireless (cellular and PCS) are included in that category. It is a matter of perspective; today, every carrier

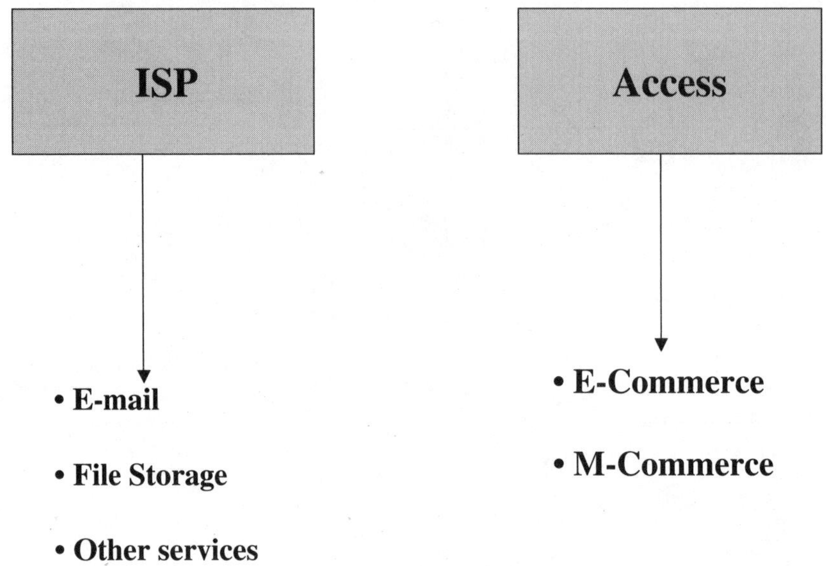

FIGURE 2-3 Business models

wants to be in the Internet business. The other carriers are perceived to be Internet "wannabees;" they are old hat and the Internet is the new game in town.

Currently, ISPs provide the following types of services and capabilities:

- E-mail
- File storage
- Web site creation
- *World Wide Web* (WWW) access
- Search engine
- Chat rooms
- Internet Portal access
- *Voice over IP* (Voice over Internet Protocol)

Figure 2-4 represents the current ISP services and capabilities.

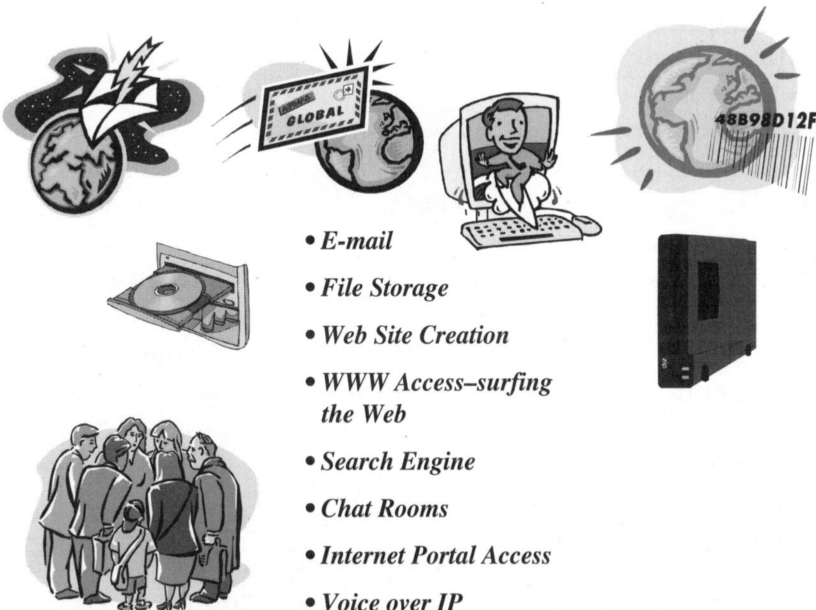

GLOBAL
48B98D12F
• E-mail
• File Storage
• Web Site Creation
• WWW Access–surfing the Web
• Search Engine
• Chat Rooms
• Internet Portal Access
• Voice over IP

FIGURE 2-4 ISP services and capabilities

E-MAIL For those in the Internet business, this may be very apparent, but for those wishing to enter the Internet community, the aforementioned listed services and capabilities may not be as obvious. In the mid-1980s, the dominant ISPs simply provided e-mail services. During the formative (technological and business) years of the Internet, very few people had a desktop computer, let alone a laptop computer. In order for the Internet business to grow, a critical mass of customers with computers had to develop first. As people began to find more uses for the computer beyond simple letter writing, more uses could be found for the Internet.

This is not to discount or minimize the importance of the e-mail service to the Internet. E-mail is to the Internet as plain old voice service is to the telephone and cellular businesses. It is the staple product of the Internet. E-mail has assumed an identity that cannot be easily equated to that of voice in the traditional telecom business (telephone and cellular). E-mail is the only service to date where a simple communiqué between family and friends may be nothing more than a joke or a poem. It is also unique in its ability to enable a user to be brief and to say nothing of substance. Not even a short voice conversation over a telephone or cellular handset can boast that kind of communication brevity. E-mail has added a whole new dimension to communication between parties. In some ways, it provides less than letter writing on paper, but in another way, provides more than letter writing of old.

E-mail enables one to carry on a conversation in near-real time. The e-mail letters enable people to engage in quick discussions about important and benign topics. Copies of letters and other documents can also be transmitted with an e-mail.

The use of instant messaging, which is possible between subscribers (customers or users) of the same ISP, enables real-time communication.

FILE STORAGE File storage refers to the storage of files that belong to ISP subscribers. File storage is performed to a limited extent today by ISPs. Files are typically stored within e-mail messages and not as separate archival data backups. Today, the majority of files that an ISP agrees to store are only stored in a temporary e-mail storage area under the user's e-mail logon name.

Some ISPs are beginning to offer archival data storage. These new ISPs offer a service that a corporate *management information system* (MIS) organization would offer to its corporate users.

WEB SITE CREATION Web site creation is a service that many ISPs offer their subscribers. The subscriber pays a small fee to use software tools that can be found online with the ISP. The software tools enable a customer to create a customized Web site with pictures, word font styles, colors, and interactive icons. This is an interesting, useful, and helpful tool as it allows the ISP's customer to take advantage of the Internet by enabling them to become more than just simple e-mail customers. This tool enables the customers to become Web information providers themselves.

Web site creation promotes the use of the Internet. Even better, the ISP becomes the host of the customer's Web site for a small fee. The ISP's computers become the physical address of the customer's Web site. The customer does not have to purchase or manage his/her own computers for the purpose of serving as a host. The Web site creation tool generates Internet usage.

Web site tools are used to create advertising. Normally, this advertising is not requested by the user, but is embedded within information or entertainment that the user has chosen to view or listen to. Advertising on the Internet is most often delivered by e-mail or in the form of banners embedded in Web pages. Advertisers normally pay the media or content providers for space or placement within the content that attracts users/customers. Interactive banners are now used to catch the attention of some customers.

WORLD WIDE WEB (WWW) ACCESS To explain *World Wide Web* (WWW) access, an explanation of the WWW is first in order. The WWW is not a single object, but a collection of Internet sites. The ISP enables a user to access multiple Internet sites. Pretend you are in orbit about the planet Earth. You look down on Earth and see thousands of Internet host computers scattered all over the planet. Figure 2-5 depicts the various hosts computers scattered throughout the Earth. If you were to connect all of the host computers using a pencil, the picture you would have drawn would look like a spider web; hence, the use of the word "web" as it relates to the Internet. The term "World Wide Web" is an accurate description of the relationship between all of the Internet sites. Figure 2-6 represents the web that has been described.

FIGURE 2-5 Internet sites all over the planet

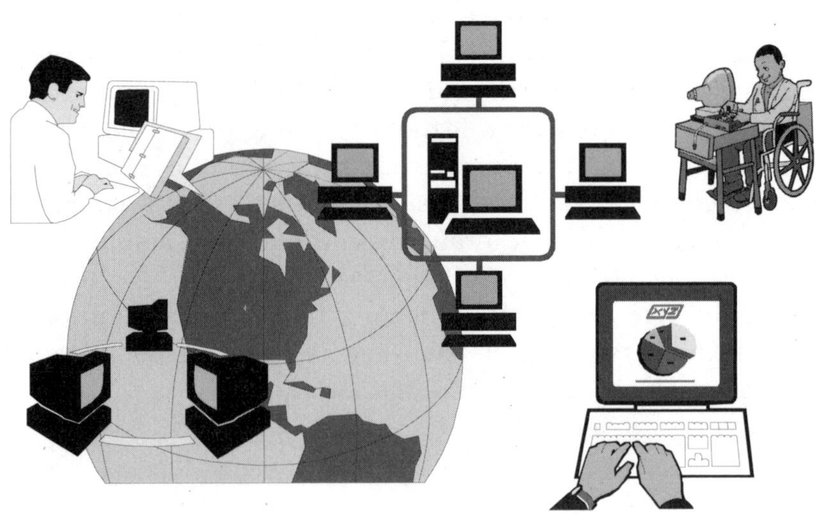

FIGURE 2-6 The World Wide Web

The carriers' ability to enable its users to communicate with other ISPs is a product of the ISP; access is a product. Users have no need to purchase service from an ISP if it cannot access the site of a company that is selling products they want. Note that the ISP may not be selling any products beyond e-mail. However, an ISP that simply offers e-mail and not the ability for its users to surf other Internet sites will not be in business for long. At first, the ISPs only provided e-mail and limited Web site access unless it hosted the Web site for that company. Note that customers had limited access to other sites during the early days of the Internet because of the limited penetration of computers and existence of commerce-related Internet sites.

Telecommunications networks, including data networks like the Internet, form webs. Today, the Internet is attempting to create a level of traffic management order within itself. The management order referred to involves the use of the "hub" concept, where single ISPs are working to act as aggregators of information for users. The hub concept, which is heavily used by other telecommunication carriers (for example, wireline and wireless carriers), has taken on the form of the "portal" in the Internet world. A hub acts as a focal point of traffic. A further discussion on the portal will appear later in this chapter.

Web access enables users to surf the ocean of information that exists at various Internet sites. Companies, libraries, schools, universities, and other institutions that have their own Internet sites are providing Websurfers information. The information is provided for a variety of reasons as shown in the following list. Some of the information is useful and some of it is harmful. The greatest challenge for parents of children is policing appropriate Internet usage.

- **Self-promotion** Advertising and marketing. Institutions are using the Internet to get users to sign up for some type of subscription or even to apply for membership in the institution (for example, a university).

- **Product sales**
- **Good will** Many institutions provide information over the Internet for the sole purpose of disseminating information. Some of this information is good and some of it is clearly immoral or illegal.

More discussion on products and services will appear in later chapters. A key point to remember is that the Internet has become a phenomenal place to obtain information or do research.

SEARCH ENGINE Search engines are a powerful attractor for Internet usage. Search engines are software-defined products that are normally provided at no charge to the subscriber (Internet user). The search engine is an application that enables the user to find information from various Web sites. The user types in a keyword(s). The search engine uses the keyword to find sites that are associated with a specific identifier. Typically, a site, which can be promoting products or viewpoints, will have specific word identifiers with which it associates (hyperlink). The unfortunate thing about keyword searches today is that the user ends up with hundreds of thousands or millions of "hits." These hits refer to the number of sites found that are associated with the keyword.

The fault does not necessarily lie with the ISP. Rather, the fault lies with the search engine. The search engines used today by the search engine companies have no way of distinguishing between a good hit or a bad hit. The following is an example of the confusion that arises when a user types in the words "red car Chrysler." One would assume that the search engine would come up with sites that list red, Chrysler cars. In fact, what appears is an enormous list of sites that address things or people that are red, have something to do with cars, or are called Chrysler. The problem is that the owners of the referenced sites can tag themselves with a variety of keywords that may have absolutely nothing to do with what the site is actually promoting. Figure 2-7 depicts what function the search engine performs.

FIGURE 2-7 The search engine–A web of site hits

Search engines lose their value when a user has to scroll through dozens, let alone hundreds or even millions of site hits. ISPs (primary product is e-mail) normally have their own limited search engines; the user can type in a keyword and it usually steers him/her to a very limited number of sites typically under the ISPs control. All ISPs provide access to hosts that provide access to search engine ISPs that provide only the search function. ISPs that provide only the search function receive their revenue from corporate advertising. The surprising thing about the e-mail ISP that had its own search engine is that as limited as it was, these ISPs had stumbled upon the correct idea. The correct idea was that the user did not want to see hundreds of thousands of hits. The e-mail ISP had stumbled upon the portal concept: controlled access.

Chat Rooms Chat rooms have been both a blessing and a curse to the Internet community. Pedophiles and other morally twisted individuals have taken to the chat room to wreak havoc on people's lives. No more needs to be said about this. On the upside, the chat room has become more than a place for people to meet and chat. It has become a place where celebrities (movie stars, television stars, famous folks, and authors) can meet fans and promote movies, books, television shows, and other products.

As a promotional tool, the chat room has brought the celebrities closer to their "public" than ever before. On television, one will see advertisements for chat room meetings with one's favorite celebrities. The chat room has become a popular tool for promoting new movies and television specials. In some cases, viewers of certain television specials are even able to meet (in chat rooms) at a specific time and day to discuss important issues that have been raised by a television special or movie. Although this form of marketing is a bit self-serving (promoting the television program or movie), it does enable an important issue to be discussed whether it is a political or social issue. To some extent, the Internet chat room has added to the social dialogue of today.

The chat room's impact on the Internet should not be minimized. Next to e-mail, the chat room is the next major product of the Internet. If one looks at how ISPs are charging for their services today, typically a flat fee is charged per month for accessing the ISP's site. It should be noted that early in an ISP's life, users were paying for e-mail. Today, the ISPs have successfully recreated themselves into being more than e-mail companies; therefore, e-mail is just one of many things an ISP sells. Chat rooms are so popular that many users obtain service from an ISP based on the technical sophistication (some chat rooms enable users to use real-time, slow motion video) of the chat room and the type of people who are in the chat room. If one logged on and entered a chat room, one might meet some fairly interesting people who also will not be physically threatening. The bottom line is that chat rooms are where strangers meet and chat.

Most ISPs have added a twist to the chat room through the provisioning of an instant messaging/writing capability. The instant messaging/writing capability enables one to write directly to a friend or family member that is also logged on to the same ISP. The great thing about this type of instant messaging is that it is a real time form of communication. An interesting aspect of this form of real time communication is that it enables users to carry on brief conversations about "nothing," that is, a casual conversation. Unlike paper letter writing of old, but very much like e-mail, instant messaging/writing assumes the characteristics of real conversations. One can have a discussion about anything from weather to one's deepest darkest secrets. In other words, one has as much chance of having a conversation about nothing as one could have about something.

Internet Portal Access The portal is a concept that became popular back in 1999. In early 1999, the IEEE published articles in a number of their societal magazines detailing research that had been performed by various computer equipment manufacturers. The issue addressed in these magazines was, "How to make surfing the Web more efficient for the user." The answer was simple and elegant: Build Internet Web site gateways. The problem with nearly all of the search engines today (and in 1999) is that the keyword search by a user would almost always result in hundreds if not millions of site listings. People do not have the time to sit and surf through that many sites. Furthermore, the searches would often end up listing sites that never had anything to do with the topic of interest.

As noted in the aforementioned sub-section on search engines, keyword searches often result in a number of site listings that simply have the keyword in question in its title or subject heading. For example, a user decides to seek an antique "Buck Rogers" toy gun that is red and made in 1930. For those who are unfamiliar with "Buck Rogers," this fictional character was a twentieth century pilot who was in suspended animation up until the twenty-fifth century. Buck Roger could be described as a combination of Indiana Jones (intelligent and resourceful), Neil Armstrong (astronaut), Charles Lindberg (intrepid pilot), Rip Van Winkle (asleep for decades), and

Batman (he fought evil aliens and other bad guys). A search for such a toy could result in a Web search listing of sites that sold toys, guns (real and toy), anything red, anything about the year 1930, anything with the word "Buck" or "Rogers," and anything "Buck Rogers." The problem or challenge for the user who made the query is that they would be forced to search the entire list. The reality is that the user would probably stop after searching a dozen or so sites. The user probably should have typed in different keywords like "antique, toy guns, Buck Rogers." However, this may not help either because the user could end up seeing sites that sold antiques (of all kinds), sites that sold toy guns (antique, new, and even pre-owned), and anything with the word "Buck," "Rogers," or "Buck Rogers." Remember that the challenge is finding the right site listed. Figure 2-8 is a pictorial of how exhaustive the list could be.

Many search engine companies claim they do not favor sites and therefore, provide objective lists. This is a laudable attitude

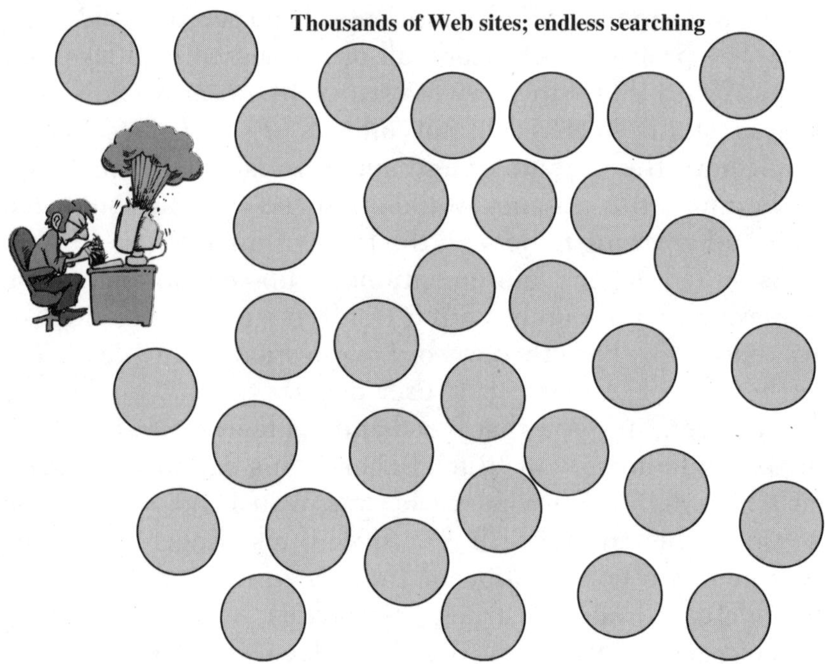
Thousands of Web sites; endless searching

FIGURE 2-8 A list of Web sites–Uncontrolled Web searches

for an industry (the Internet) that is the home of open information access. Unfortunately, from a practical standpoint, too much of anything can be bad. Users of the Internet should be treated like customers; their time and money are valuable. Without the user, the Internet companies have no business. The portal is an organized way of facilitating the users' searches. It should be viewed as controlled and managed information access. Portals are designed to focus on specific needs. For example, a children's portal may focus on new toys being marketed this year. Another portal may focus on sports cars made in the United States between the years 1950 through 1999. Another portal may focus on charities in a given geographic area. Another portal may focus on specific medical issues. One can see that portals can be and are established for a variety of products, issues, and activities. Figure 2-9 illustrates this point regarding what portals support.

The portals are managed in such a way that one can visualize the portal-Web site relationship as an organization chart where specific positions are related to other positions (simply

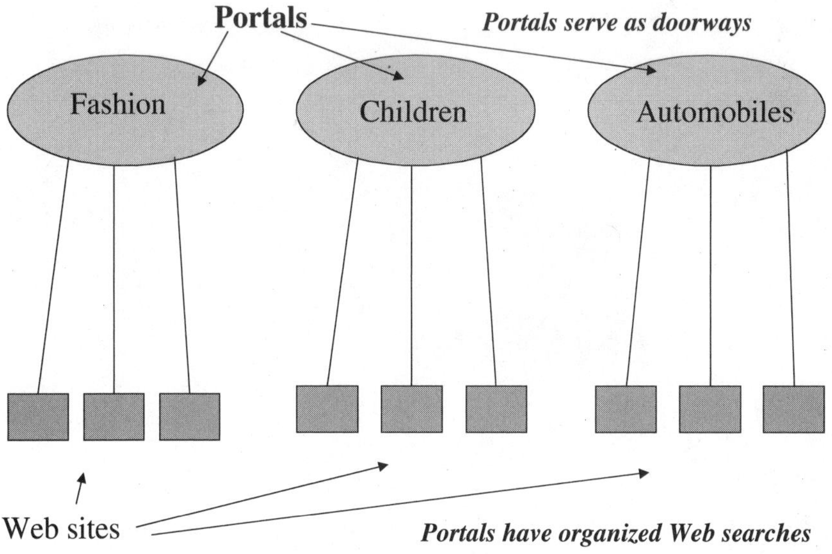

FIGURE 2-9 Portals support a variety of products, issues, and activities

follow the lines of organization). Another way of viewing a portal is in the same way as viewing a highway toll both. As multiple lanes of vehicular traffic approach tolls booths, the cars are often forced to merge into fewer lanes. In telecommunications terms, this is called traffic aggregation. Tandem switches aggregate traffic from multiple points into itself for the purpose of redistributing traffic into the local network. Portals are essentially points of search and information access aggregation. Figure 2-10 is an illustration of the portal concept.

The portal is an excellent application of basic traffic management, however, downsides exist for the user. The portal is managed access; therefore, one has to trust that the portal manager has established links to the appropriate sites, otherwise the user will be forced to search through hundreds of sites. This is a challenge for the portal manager.

Portal management is a new opportunity for Internet entrepreneurs. Rather than providing content on the Web, the portal manager can collect a fee for obtaining information hits on customer sites. This type of portal business model has its weak

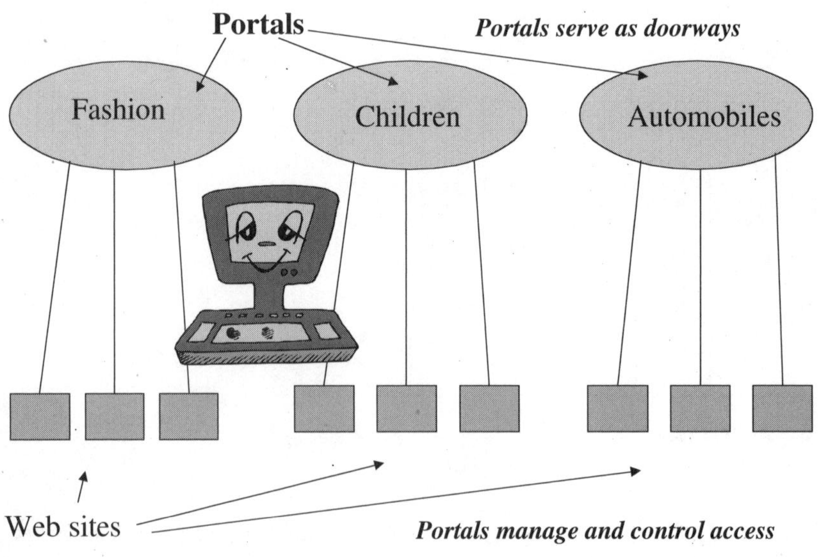

FIGURE 2-10 The portal

spots because managing information access does not necessarily mean any intellectual property enables the portal manager to differentiate itself from other portal managers. The newness of the portal has captured the imagination of investors and technologists. The ability to control the flow of information is a key function of any service provider (both non-Internet and Internet carriers). Both the business and technical communities have yet to fully explore the opportunities behind the portal. Figure 2-11 illustrates the control aspect of the portal.

The portal manager can be considered an *application service provider* (ASP). The ASP is a new term that the Internet industry is still defining within itself. In the traditional voice telecom world, the ASP is the software and hardware vendor who is selling software and hardware that is connected to the carriers' main switching system. Typically, ASPs are not the same vendors as the ones who provide the main switch or infrastructure. ASPs in the Internet world sell services to users; however, the ASPs are not the ISPs or other type of Internet carrier. These non-Internet ASPs provide services and capabilities such as:

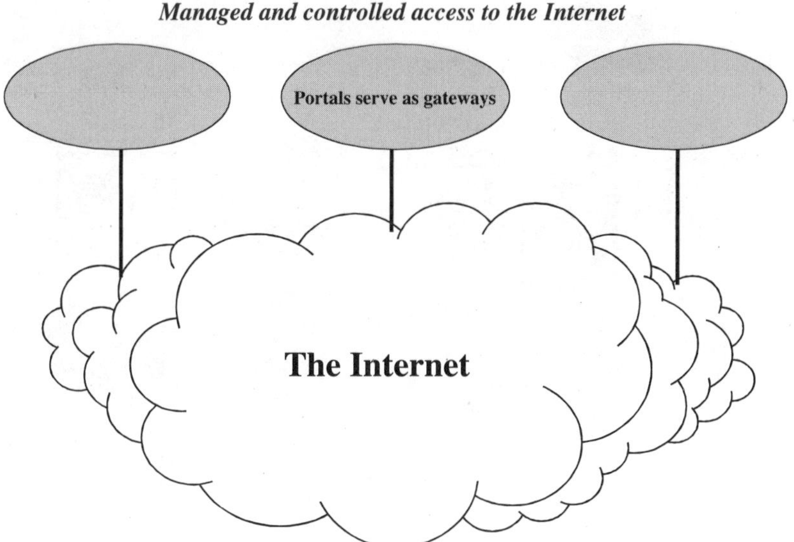

FIGURE 2-11 The portal is an information access gateway

- **Database Management** For instance, subscriber profile storage. The *Service Control Point* (SCP), the *Home Location Register* (HLR), and *Visitor Location register* (VLR) are examples of outboard systems that can be attached to the main switch.
- **Voice Mail**
- **Fax Mail**
- **Wireless finger printing**
- **Service bureaus** As noted in *Telecommunications Internetworking*, the service bureaus provide a variety of needs to the carrier.

Figure 2-12 is a representation of how non-Internet ASPs work with non-Internet carriers.

ASPs will be discussed in greater detail later in the book. The point to keep in mind is that the industry that services Internet carriers is growing and blossoming quickly.

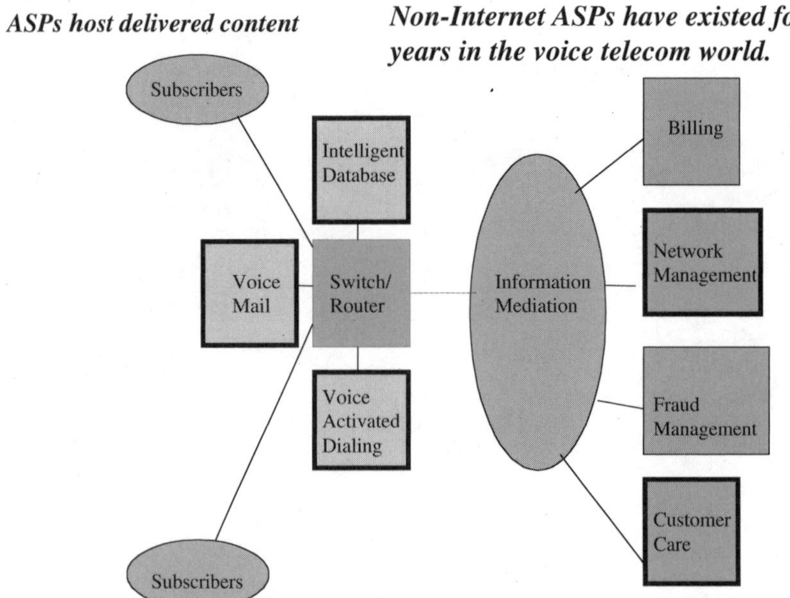

FIGURE 2-12 The non-Internet ASP

VOICE OVER IP (VOICE OVER INTERNET PROTOCOL) Voice service has the broadest appeal to the marketplace. The voice service is the lowest common denominator in the telecommunications service provisioning equation. In other words, it defines what all customers are looking for. Internet companies/carriers realize this for they are all seeking to look like the telephone company by providing voice. One should remember that not everyone is using the Internet. The power of the wireline telephone company is its familiarity with all people. Voice is the telephone company's primary service. Although one could easily make a technological case stating that the telephone company is an information processing carrier, voice is the big service to the non-technical user (which is the majority of the marketplace).

The challenge for the ISP is a technological one. Voice and video can be provided over the Internet today; however, the protocol today, known as both IP and *IP version 4* (Ipv4), has several flaws or perceived flaws. The flaws or challenges are related to the following:

- Reliability
- Security
- Service quality
- Guaranteed service levels
- Addressing schemes

The *Internet Protocol* (IP), currently IPv4, is not at all well-suited for time sensitive (delay sensitive) information. Voice and video is time sensitive information. Note that time delays disrupt the sound and look of voice and video transmissions. Disruption of quality can be so extensive that the distortions in voice quality and video quality are unacceptable annoyances. ISPs wishing to become the primary provider of telecommunications services to the marketplace will need to provide voice as a service and work toward the one quality benchmark that the entire marketplace knows: the telephone company.

More information on VoIP will appear later in the book. Figure 2-13 is a representation of voice in the non-Internet and Internet worlds.

ACCESS: COMMERCE MODELS

The Internet is an information medium that enables users to access many different databases. The search engines employed in the industry facilitate users' search capabilities. Buying products over the Internet from a retailer or wholesaler requires one to access the database of the seller. Sending and receiving e-mail is also an attribute of access because the users allow others to gain access to them via this messaging service. Access is a capability that lies at the heart of the Internet business. This access capability is what makes the Internet a valuable tool.

A child or adult can quickly obtain information from a library on the availability of a book, find out airline flight schedules, train schedules, movie time schedules, and information on almost any topic. The information obtained may be used for a

- **Reliability**
- **Security**
- **Service Quality**
- **Guaranteed Service Levels**
- **Addressing Schemes**

Voice will be the lowest common denominator for the ISP.

In the minds of the customer, voice is the most basic service need.

FIGURE 2-13 Voice: Common Issues

variety of purposes including purchasing products from the database holder. The user is limited by what the targeted database will allow. The user is also limited by the software on either the desktop computer or by what the targeted host site will allow. The purchasing of products is an activity that has come to be known as *e-commerce* (Electronic Commerce). E-commerce is normally seen as catalog purchasing off of the Internet, where the selling party displays pictures of its products, interactive (point and click) icons, and provides online purchase order forms. These product sellers even have their databases tied into credit card validation databases. Figure 2-14 illustrates access and its fundamental importance to the Internet.

E-commerce is an activity/application and not a product to be sold to an Internet user. However, it is an activity/application that has captured the interests of both the commerce and investment sides of the business community. E-commerce has given birth to a number of different business relationships over the Internet:

FIGURE 2-14 Internet and access

Business-to-Business (B2B) The B2B relationship describes the relationship between two or more businesses selling services to one another. Suppliers of small parts to an automobile manufacturer like General Motors would be one example of a B2B relationship. Another example would be the supplier of pencil erasers to the pencil manufacturer. One can take the supplier of ink to the newspaper publisher or to the magazine publisher (news is still supplied to people on paper). Today, the sub-division of relationships has also resulted in the *Business-to-Small Business* relationship, B2SB.

Business-to-Consumer (B2C) The B2C relationship is the one with which the average consumer is familiar: retailers selling products online. Consider this online catalogue shopping.

Business-to-Government (B2G) The B2G relationship is a new one that is growing over the Internet. This relationship is about the business community doing business with sectors of the government. The government, whether local, state, national, or international, is the single largest customer of goods and services. To many in the telecommunications business, the government is the single largest customer of telecommunications goods.

Exchange-to-Exchange (X2X) The X2X relationship is the linkage of multiple B2B relationships/marketplaces. X2X e-commerce is the next logical step beyond B2B or B2SB. Companies are able to purchase supplies/products at competitive prices. X2X can be likened to a farmers' market, where large food distribution companies bid for farmers' products. Manufacturers and vendors of all kinds seek to purchase supplies and services over the Internet. Steel is an example of a product that was often sold between companies and countries via a network of dealers. Now, steel can be bought and sold over an Internet based exchange. X2X e-commerce could ultimately result in lower prices to the consumer because of the lower costs to the retail outlets and to the distributors (that is, the middleman).

Figures 2-15, 2-16, 2-17, and 2-18 are illustrations of the various business relationships.

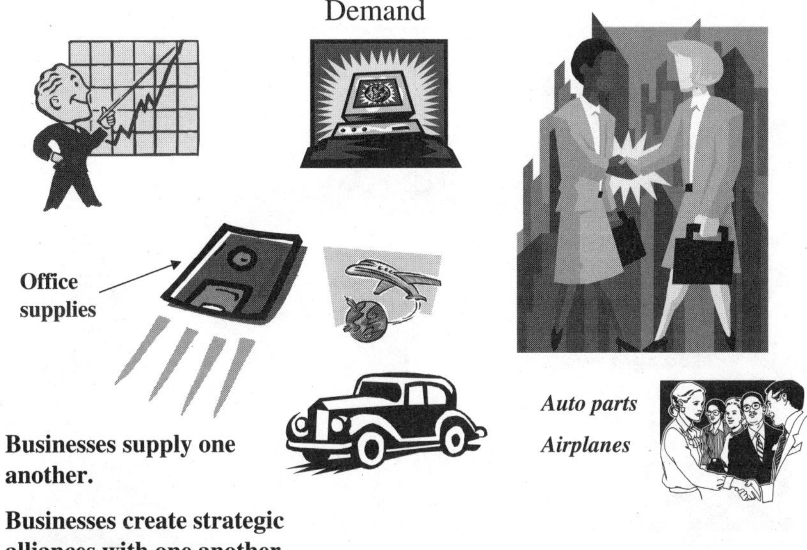
Demand
Office supplies
Auto parts
Airplanes
Businesses supply one another.
Businesses create strategic alliances with one another.

FIGURE 2-15 Business-to-Business (B2B)

BEEP!
BEEP!
BOOKSTORE
VOICE
MAIL
Businesses sell goods to consumers.

FIGURE 2-16 Business-to-Consumer (B2C)

FIGURE 2-17 Business-to-Government (B2G)

Selling goods in bulk on the open market.

FIGURE 2-18 Exchange-to-Exchange (X2X)

Actually, these relationships existed before the Internet. The names just look modern and cool. However, the Internet has been able to support and even nurture these relationships. The ability to communicate information between various businesses can either expand some businesses or put others out of business. The Internet marketplace has become an integral component of commerce. Many of the related sites are serving as portals for members of the site.

A recent twist to the B2C marketplace was a Web site I recently saw that supported the carpenter business. I was recently seeking carpenters willing to work in my hometown. The town has a nearly crippling number of building codes that frighten off contractors. I was able to access the Web site and was given a list of carpenters who expressed interest in working in my town. This Web site reduced my search by a thousand fold; rather than searching through a telephone book, I was able to enter my parameters and other related project information and received an answer within minutes. A county-based trade association administered this carpenter Web site. It can only be assumed that the trade association members pay for this service. This does not mean I saw every contractor willing to work on my home because some of the members may not have wanted their names listed on the Web site. However, it is important that my search time was reduced. The fact is that most consumers seeking contractor estimates will only call about three or four companies at most for the job cost estimates.

E-marketplaces are still growing. It should be noted that the Internet has open and established relationships between companies and consumers as never before. Nearly instantaneous price quotes, inventory queries, sales discounts, and order placements can be made over the Internet. The ability of the Internet to provide accurate information is limited by whatever data is made available to the users. In the end, it is all about access. At some point in the near future, the Internet will be the most vital component of doing business. Figure 2-19 is an illustration emphasizing the role of the Internet in the marketplace.

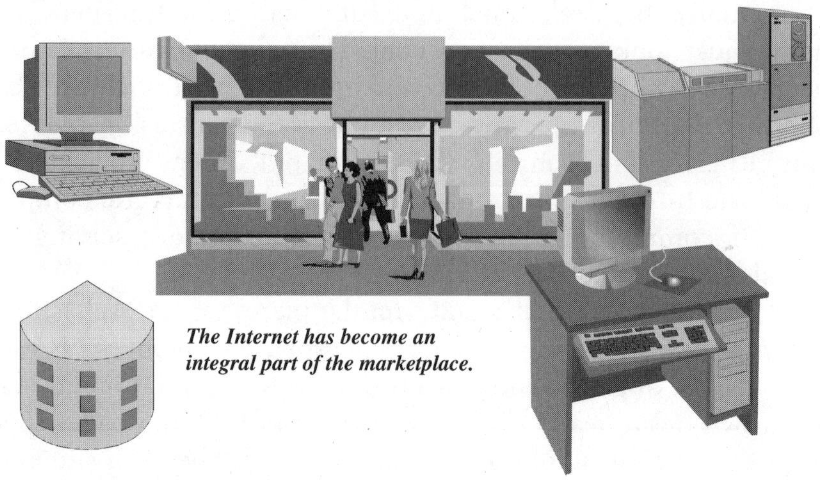
The Internet has become an
integral part of the marketplace.

FIGURE 2-19 The marketplace and the Internet

Although it may sound facetious, a potential area of growth is the *Taxpayer-to-Government* (T2G) relationship. Today, taxpayers can file their income taxes over the Internet. They can find a variety of information from the various government databases. Furthermore, taxpayers can file any number of requests for information (from government agency) over the Internet. It is unlikely that the government will use the Internet as its primary way of conducting business with the taxpayer because the government can never be faceless, but it does enhance the flow of communication between the government and the people it serves. Figure 2-20 illustrates the T2G concept.

Another way of looking at the various e-marketplace relationships is by simply categorizing all of the previous relationships as *Information/Product-to-Consumer* (IP2C) relationships. All of the aforementioned relationships have one common factor; they send information or product to someone else. In any business relationship, someone is the consumer or customer. Figure 2-21 illustrates the IP2C concept.

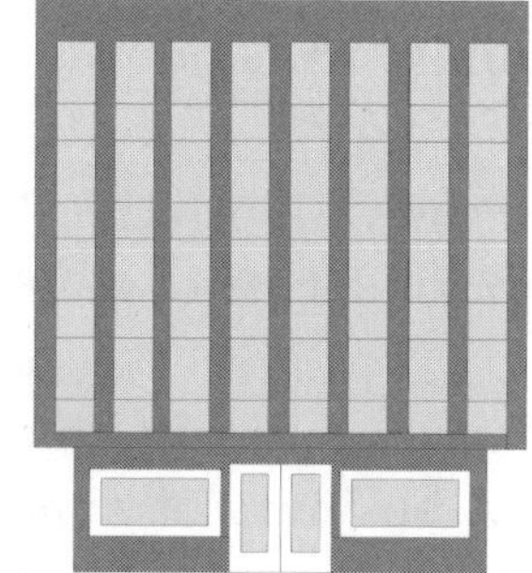

• Filing taxes
• Requesting information
• Nationwide traffic safety alerts
• Filing reports to the government
• Communicating over the Internet

FIGURE 2-20 T2G

FIGURE 2-21 IP2C

The two broad categories can be further broadened as being one of or a combination of both:

- **Service Provisioning** ISPs provide service to the mass market.
- **Access** ISPs and the various other Internet players are all providing access to some type of database.

This short list is not meant to give the reader the impression that the Internet is a business of simplicity. However, these views are meant to provide a way of understanding the Internet business. The Internet is a new medium for doing business. The rules for conducting business do not change. The business community is still working on how to create opportunities for the Internet.

Basic micro- and macro-economic rules have not changed. The rules of supply and demand are still underlying principles of the sale of products. The customer still pays the bills. All of the aforementioned models and sub-categories have been built around what people already know and understand. Some believe that the Internet has created a new kind of economy called e-economy or e-conomy. The recent stock market collapse of the Internet stocks is proof that rules of doing business over the Internet or not over the Internet are the same. Figure 2-22 is a representation of the e-conomy/e-economy.

THE EVOLVING BUSINESS MODEL

Up to this point, this book has described where the Internet industry has gone to date. The question remains, where is the Internet business going? The answer to the question is simple. The Internet business model is at the beginning of its evolution and will evolve into anything the marketplace will drive it toward. The marketplace will drive the Internet's technological development down a path that is guided by business/market

FIGURE 2-22 E-conomy/E-economy

parameters familiar to all of us. The challenge for the Internet players is to find that niche or mass market idea that will generate revenue for themselves while providing a useful service. A guiding principle for the Internet player is to first look at how things are done today and then look at how the Internet can add value, which is a very basic way of approaching any matter.

Today, the Internet space is filled with companies all claiming to provide some kind of Internet service. Some of these company types are as follows:

- ISPs
- Gateways
- ASPs
- Content providers

In the end, all companies claim to have some control over the access to the customer and provide some kind of service to the customer.

The marketplace will find a multitude of ways to find uses for the Internet. As an information medium that is in its infancy, thousands of opportunities are available for a system that supports the manipulation and transmission of information. The Internet's single greatest attribute is its ability to facilitate access. By focusing on the access aspect of the medium, investors and entrepreneurs will be able to find new opportunities. The reader should remember the old axiom: Information is power. The Internet enables everyone, both the individual and the large corporation, to manage the flow of information or product between parties. This is called "access," and this is ultimately "power."

The marketplace is always seeking new ways to improve the quality of life by making things:

- Easier to do.
- Easier to find.
- Faster to do.
- Less expensive to do.
- Quicker to travel to.
- More fun to do.
- More interesting to perform.
- More interesting and higher in quality to view.
- More interesting and higher in quality to listen to.

Figure 2-23 illustrates the aforementioned points about access and the basic goals of the marketplace.

The parameters that were just described impact all products that are sold or created whether they are over the Internet or not. People will not deviate from these parameters. To a great extent, e-commerce has done an excellent job in meeting the marketplace's basic goals. However, people must see continuing value in the Internet approach, otherwise no financial value exists to the provider. People like doing business with those companies that provide continuing value rather than one time

Why the Internet?

- **Easier**
- **Faster**
- **Less Expensive**
- **More Fun**
- **Better Looking**

FIGURE 2-23 Goals of the marketplace

discount sales. The markets can be broken down and combined in any number of ways; however, the following are broad categories:

- Home
- Office
- Travel for business
- Travel for vacation
- Entertainment for the home
- Entertainment in publicly accessible areas such as amusement parks and movie theaters
- Education at home, in the school, museums

The Internet will evolve down two paths: technological and business. The technology of the Internet will change in order to meet the needs of the marketplace. The business of the Internet will evolve in order to ensure the generation of revenue. Technology and business work hand in hand to achieve the same goal: creating value in our lives.

The future of the Internet will depend on how both technology and business work together to achieve value. Focusing on access will enable all Internet industry players to find and create opportunities. These new opportunities will be focused around access. Figure 2-24 re-emphasizes the importance of access in the world of the Internet.

SUMMARY

In the late 1990s, people were amazed and bewildered by the growth of the Internet. A number of things worked in favor of the Internet industry's growth and it all took place over a period of nearly 40 years. These factors and innovations occurred completely independent of one another and were all initially driven by different reasons:

The mainframe computer Faster way of performing business (accounting) and scientific calculations.

The ARPANET A way of ensuring more secure national security communications.

The desktop computer First bringing games to the home (remember the first Atari, the Texas Instrument 99-4A, and the Commodore) and then improving the workplace (the word processor increased the productivity of the steno pool and the secretary).

The laptop computer Facilitated the usefulness of the computer for the businessperson. This usefulness first took the form of letter writing by the worker thereby reducing the need for secretaries, and increasing the timeliness of letters by eliminating one person in the process.

Local area network (LAN) concept Linking a mainframe electronically.

The Ethernet Linking the LANs.

FIGURE 2-24 Access and the Internet

Netscape Netscape the computer industry's first (and innovative) and successful attempt to find a useful way of speaking to the various educational research Web sites. This first search engine gave the Internet weight for research of all kinds, both educational and commercial.

E-mail E-mail opened the Internet to the mass market.

The industry was built upon a variety of innovations and initiatives. The entire telecommunications industry like any industry is built upon the work and creativity of various people and other industries. All of the previously mentioned innovations and factors had one thing in common: access. Access is about moving information and product between people. The

Internet, both wireless and wireline, has become the primary means of accessing the home and business. Figure 2-25 is another depiction of the Internet business.

The next chapter will address the technical and business challenges of the future ISP.

BEEP!
BEEP!
Every telecom player is entering the Internet space
Who Occupies The Internet
Business Space?

FIGURE 2-25 Who occupies the Internet business space?

CHAPTER 3

CHALLENGES OF THE INTERNET BUSINESS—BUSINESS AND TECHNOLOGY

The Internet has been a haven for investment money, entrepreneurs, and established companies seeking new markets. The Internet has gone from a text-based, e-mail only medium to an interactive, multi-media communication medium. However, a great deal of work is needed on the technical side in order to fully explore the multi-media business opportunities. The reader should understand that the information in this chapter is applicable to the wireless and wireline Internet industries. The business challenges of the Internet industry are as daunting as the technical challenges. The challenges can be divided into two categories: technical and business.

BUSINESS CHALLENGES

The danger of doing business over the Internet is that people like speaking with other people. Recently, a large toy company had a disastrous circumstance befall itself when it could not fill toy orders placed through its own Internet site. The toy company not only failed to fill the orders, but also could not address customer complaints over missing orders and unfilled orders for toys. Customers became so angry that they refused the toys when they finally did arrive and as a result, many people no longer shop online with this company. Many of these people also refuse to go to the brick and mortar stores of this company. The Internet can be impersonal. Before the Internet, companies lost customers who felt they had been mistreated by customer care personnel. Today, in a world filled with ISPs and retailers with their own Web sites, customers still want to speak with a human when a problem arises. It should be noted that other service providers like wireless and wireline carriers have received numerous complaints when their automated directory assistance, repair service, and operator service centers failed to respond quickly or adequately enough to the customer.

ISPs may believe that their customer base is more sophisticated and more tolerant of automated/e-mail driven customer care systems. This view holds true only for the high-tech users who tend to be first or early adopters of technology. In reality, if an individual walks into any brick and mortar store for electronics (retail music and video equipment or computers) and watches the customer care line, he/she will see customers of all kinds shouting at salespeople and store managers. People who spend money want to be treated like "kings" and "queens." The provider of service wants the customer's money, so he/she must learn quickly the most basic of all business credos, "The customer is always right." This is not some old and outdated idea. If an individual watches what happens when a customer does not like the attitude of a salesperson, he/she will see how a person can go from calm to mad in 60 seconds flat. Figure 3-1 is an illustration of the customer relationship.

Retailers doing business over the Internet have the same problems with inventory as the brick and mortar stores. When a customer cannot find a product in a brick and mortar store, they simply look elsewhere. The Internet has set the expectations of the consumer higher than if they were going to a shop-

- **Products**
- **Services**
- **Online Shopping**

The customer generates the income for internet business.

FIGURE 3-1 The customer is king/queen

ping mall looking for a product. To many people, the word "Internet" has come to mean "instantaneous." From a business/commerce perspective, the consumer expects instantaneous service over the Internet. Stores cannot be expected to maintain an unlimited inventory for online shoppers. This is a basic rule of supply and demand. However, the consumer does not care about the store's accounting issues. The Internet opened up the marketplace for retailers. Where once a retailer would never have been doing business with a person in a particular location because a brick and mortar store was not present, the Internet now has made it possible for the retailer to do business worldwide. However, the rules of business and customer care have not changed. This point will be discussed futher, later in this section. Figure 3-2 is a depiction comparing the Internet and brick and mortar stores.

How do you express your frustration and anger over the Internet?

The brick and mortar store enables customers to interact with the store in person.

Even speaking to customer care over the telephone is more desirable than Internet communication.

FIGURE 3-2 Expectations: The Internet and brick and mortar stores

POSITIVE AND NEGATIVE EFFECTS OF THE INTERNET

The positive effects of the Internet on the retail marketplace and business community in general have been:

Greater access to new geographic markets Enables new places and people to sell products.

Facilitation and speed of communication between parties Instead of spending hours and days writing and sending memos, e-mail provides real time communication. The mobile telephone and wireline telephone enable real time communication, but e-mail enables one to send large documents along with the message. E-mail has sped up and facilitated communication of complex issues. Group communication is especially made easier. Advertising is also made easier, and even interactive.

E-mail has enabled many people to discuss issues without anger or emotion. Often times issues cannot be resolved when the parties involved in a verbal discussion are emotional. E-mail can bring professionalism to an otherwise charged situation. However, e-mail communication is only as effective as the communicator's ability to articulate their views.

The negative effects of the Internet on the retail marketplace and business community in general have been:

Greater access to new geographic markets Greater access to new markets means greater opportunity for not meeting the demands of more people.

Facilitation and speed of communication between parties In the business world, when once writing a memo on paper meant taking time and therefore facilitating careful thought, writing a message (e-mail) out of anger comes easier and faster over the Internet. The Internet has wreaked havoc in some business offices. Some employees lose their sense of reason and write memos filled with anger towards superiors and co-workers.

One example occurred not to long ago with a colleague of mine. An employee received a memo from my friend, praising the employee for the fine job this person had performed all year. As this person's organizational superior, it was incumbent upon her to send this note of praise. This person was considered the organization's top performer. Unfortunately, when this person had received the memo, she had apparently had a rough week in the office. Rather than simply accepting the memo without comment, she responded to her own office mate via e-mail with cynicism and disdain for "corporate officers who claimed they knew of their fine work and wished to express their gratitude for the person's work via a bonus." This employee went so far as to say to her friend that my friend would not know them from "a hole in the wall." What the employee did not realize until too late was that out of anger, she apparently hit the "reply all" icon and sent my friend the message, and not her friend. My friend did not respond directly to the note. This person apparently realized (too late) what had happened via a conversation with the originally intended party of the "cynical response" memo. The person who the memo was intended for voluntarily approached my friend and claimed that they did not put this other employee up to writing the first memo. My friend did not approach her direct reports or their direct reports about the memo; however, she did walk over to this employee in the company cafeteria and introduced himself. My friend told me the employee was so flustered that they could not speak. My friend advised the employee that as tough as some weeks are, they should always maintain self-control: "Take a deep breath and count to ten before you do anything rash again." In her opinion, e-mail made it simple for people to simply express their anger.

E-mail lends itself to brevity. As a result, some employees use the tool as a way of communicating messages that seem ill-thought out. It has also become an outlet for passing jokes (some tolerable and others inappropriate for the workplace) and other inappropriate material (pornography).

An unfortunate aspect is that e-mail has become a replacement for voice conversations and paper written memos (including good manners). It is much easier to write an e-mail than to deal with emotional human beings who are either not thinking clearly and are excited or are simply being difficult and obnoxious. Further, some folks use e-mail just to be plain rude and obstructive. However, at some point, people will need to speak to one another. Voice conversation cannot be replaced entirely.

Higher expectations Many consumers have come to believe that the Internet means faster and better, which is not necessarily true. Unfortunately, the Internet has caused retailers to expend enormous time, energy, and resources trying to meet customer requests. Furthermore, companies that use the Internet to communicate to the marketplace have found that maintaining their Web sites requires a full-time staff dedicated to the Web site.

Figure 3-3 illustrates the Internet's pros and cons.

- *Access to all places and people*
- *Facilitating communications*

- *Overstepping capabilities of Internet retailer*
- *Too much communication*
- *Brevity in communication Meaningless communiques*
- *Higher customer expectations*

FIGURE 3-3 Pros and cons

Pros and cons always exist in any business. It is important to be aware of those aspects of the business that can negatively impact the customer. The challenges of any business, including the Internet, are technical and business-related. Knowing what you want to sell, how you want to sell it, where you want to sell it, how to make the product, and where to obtain the resources to make the product are among the questions that one needs to address in order to start and run a business. The ultimate challenge of any business is having an effective plan. The next section looks at the basics of running a business.

BASICS OF RUNNING A BUSINESS

In the year 2000, investors have come to their senses and realized that even though the Internet opened up new marketplaces, it did not change the rules of economics, financing, accounting, inventory control, debt analysis, or equity analysis.

One cannot overstress the need to understand what the Internet is as a business medium. During the early days of the Internet gold rush, it appeared that Wall Street and investors were not looking deeply into the business plans and organizations of new startups. It was easy just to invest because the rush for investment opportunities could easily find another "Microsoft." Many banks and investors failed to ask the right questions and look for basic mechanisms and management team skill sets.

The following are basic questions and issues one needs to address in order to run a business:

- Know what you are selling.
- Understand why anyone would want to buy your product or surf your Web site.
- What makes you think they will buy from you? Your Internet Web site could be so hard to surf that the user simply goes to another site.
- Know how big your market is.
- How much of this market do you expect to capture for your company?
- Why do you think you will be successful?
- What does your team look like? (In other words, who are your management team members?)
- What financial control mechanisms are in place? Such mechanisms would include things like inventory control, expense travel allowance rules, equipment purchases, signature level requirements, managing benefit package costs to the company, creative ways of not paying taxes, and identifying and quantifying operations costs.
- What are the company's hiring practices?
- What customer care processes are in place?

Plan one's Internet-related business follows the form of

standard product or business planning practices. The planning cycle for any endeavor is the same for effort in all disciplines of science, engineering, development, or business. The terminology may differ, but the basics of logical planning are the same. Figure 3-4 illustrates the basic planning process as described in the aforementioned list.

Planning the Business: The Basics

Planning is a structured activity. Planning takes the form of project management, the scientific method of experimentation, medical diagnosis, financial planning, career planning, planning for the future of one's children, planning home finances, and so on. Planning is a structured activity that establishes a course of action. It starts with a desire for a result and hopefully ends with the desired result. Those entering the Internet business or an Internet player entering the mass-market voice/vertical services telecom business should not be intimidated with

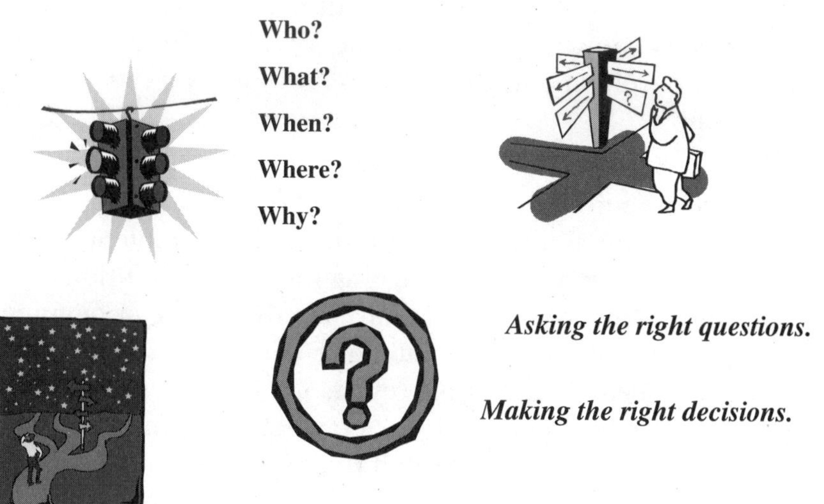

Who?
What?
When?
Where?
Why?
Asking the right questions.
Making the right decisions.

FIGURE 3-4 Who, what, when, where, and why: Asking the right questions.

the business segment that they are looking at because every opportunity can be approached logically. Planning for any endeavor involves five aspects:

Problem/Goal Identification What is the problem you are trying to solve? What goal are you trying to accomplish?

Problem/Goal Analysis Gathering the necessary data needed to take that first step to ensure movement forward on one's action plan. The data needed to create a solution may include customer requirements, data about competitors' solutions, or data about current practices. More importantly, this step is needed to identify what information one does not know. Recognizing one's limitations is essential to success.

Solution Design Designing a solution to attain one's goals or resolving the problem to one's satisfaction. This step can include writing software, policies, or procedures, or designing the hardware or building.

Implementation/Action Identify the steps and people needed to act on the proposed solution. For many people, this step is known simply as the action plan. To those in product development, this step includes testing the solution.

Ongoing Operations and Maintenance Any business or endeavor requires the knowledge and understanding of what will happen after the solution has been implemented. Does the solution have an ongoing life? Is the solution a one-time event that does not have any recurring or ongoing role? Every story has some kind of ending. In a scientific or medical endeavor, the effort results in either a published paper or perhaps a cure. However, what happens after the solution has been implemented the first time? In a business, this step means identifying all of the necessary resources to continue the business. The necessary resources of a business can include equipment, land, people, financing, software, an organizational structure that supports the business, marketing, a viable market segment, and so on.

Figure 3-5 is another illustration of the planning process.

Each of the aforementioned steps can be further subdivided into other steps or phases. The Internet player can view entering the telecom business as a planning effort similar to the one used to create their original Internet endeavor. For those seeking to enter the Internet business, the planning process is the same. It is important to keep things in perspective. As different as the traditional wireline and wireless telecommunications businesses appear to be from the Internet business, they are very much the same. Both the non-Internet and Internet industries work to serve a customer. The customer is the most important person to the non-Internet and Internet companies. Serving the customer is the common goal for both telecommunications segments. Planning for this customer service requires the same logical thought process. Figure 3-6 illustrates the commonality in the non-Internet and Internet communities.

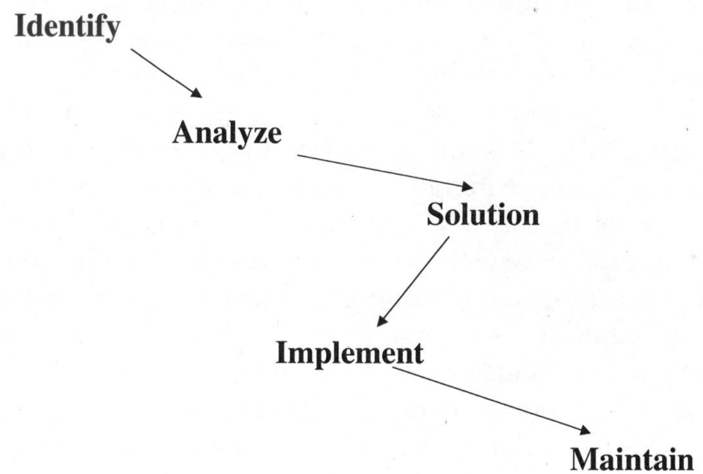

FIGURE 3-5 The planning process

FIGURE 3-6 Serving the customer: The common goal

THE MOST IMPORTANT PART OF THE BUSINESS: THE CUSTOMER

The customer is an important focus for any company. Surprisingly, the customer is often the most overlooked participant in the business community. Companies can easily internalize their views about their product and forget the customer in the process. Many companies internalize their market views of their product and essentially end up believing in their own marketing hype. It is important for companies to support their product and believe in their product's ability to have value; however, a point is drawn where people can stop listening to the customer and only listen to themselves. Companies that have been in business and are doing well usually use a combination of salespeople, product marketing folks, and focus groups to gather data and updates about the customer. The incident mentioned earlier in this chapter concerning the toy company leaps out as the most glaring example of the retail Internet marketplace going wrong.

During the early days of the Internet, ISPs and other Internet users could easily identify their customers. These customers were the "early adopters." Early adopters are customers or users of products who tend to like to be either the first on the "block" to own or do something or those who love to play with the newest gadget on the market. Early adopters are brave by comparison to non-early adopters. The early adopter market is smaller than the mass market, making it easier to identify ways in which to satisfy the market. Figure 3-7 is a depiction of the early adopter versus the mass market.

Internet retailers need to be observant when their customer base shifts from the early adopters to the mass market. When the shift to the mass market occurs, it is important to shift one's marketing programs and effort so that the entire base is being advertised to or communicated with. The mass market shift may occur independent of anything the retailer is doing in terms of marketing itself. The shift may have occurred simply

FIGURE 3-7 Early adopter versus the mass market

because of a market ground swell for the product type. The Internet is an interesting study from a consumer market perspective. First, the overall customer base needed to become Internet conversant. Second, customer segments within the entire customer base needed to find their shopping Web site. Surprisingly, the Internet industry is at a point where many retailers are not even online. Therefore, customers are looking but not finding. As time passes, increasingly more members of the retail community will go online. At some point in the distant future, every retailer will be online. However, it seems unlikely that every retailer will be online unless an ISP works with the small business community. For now, the quick money is in the large business community.

In order to plan a retail business, ISP business, Internet back office support (equipment suppliers), or an e-commerce-related site, one needs to apply standard business planning practices. The first practice is identifying with which market to do business. The next is to understand the customer segment's needs. Identifying the market to do business with is easy. Understanding what the customer is seeking is more difficult. As noted earlier, companies often internalize their views of the market; this is typically called "looking into the mirror." Looking into the mirror has its benefits if someone takes an honest look at his or herself.

ANALYZING CUSTOMER NEEDS

A variety of analytical tools are at the disposal of a company. It is important that the correct tool is used and applied properly. My preferences for marketing tools are two types of market analyses that an ISP can perform. The first is a method of competitive analysis and the second is called a customer satisfaction survey. Performing a structured analysis of the marketplace is a necessary administrative evil. During the late 1990s, hundreds of ISPs leaped into the telecommunications community at an alarming rate without carefully looking at the marketplace. Performing a series of market analyses will not guarantee success for any business, but without a structured review of the market, the ISP (or any company for that matter) runs the risk of failing.

The competitive analysis method referred to here is known as SWOT. SWOT is shorthand for Strengths, Weaknesses, Opportunities, and Threats. The SWOT analysis reviews the company from an internalized viewpoint. SWOT examines the company's *strengths* (S), *weaknesses* (W), potential market *opportunities* (O), and external *threats* (T). The purpose of such an analysis is to make a company take a close and honest look at itself. The company performing a SWOT examines its own strengths, which could be software writing, hardware development, or Web page creation. Once those strengths are identified, the company then examines its own weaknesses. These weaknesses could be in areas where the company should be strong. The opportunity review is an examination of the opportunities in the marketplace for the company's product. The examination could result in identifying opportunities that had previously gone unnoticed. The opportunity analysis may even result in identifying opportunities for products that are still on the drawing boards. The threat analysis examines the company's external threats. The external threats may include other competitors, national or global economic forces, supply shortages, or even shifts in customer needs.

A SWOT is not simply an identification process; it is an analysis that is coupled with a plan. The plan is strategic and tactical in nature. The plan must also outline a specific course of action. Figure 3-8 for an outline of this plan.

Customer surveys are often the last marketing tool a company views. Companies often believe that customer surveys are simply tools that justify one's position. Unfortunately, many surveys are worded in such a way that they tend to skew a respondent's views in a way that justifies a company's existing position. Nevertheless, if the survey is properly constructed, an unbiased set of answers and valuable information can be provided to the company. The marketing process is a process by which people obtain what they need and desire through creating, offering,

Strengths – Your own company's strengths

Weaknesses – Your own company's weaknesses

Opportunities – Examining the marketplace for opportunities

Threats – A competitive analysis and external forces review

Analyzing the competitive market by looking at the environment, starting with an internal look at oneself and then looking outward.

FIGURE 3-8 SWOT

and exchanging products with other people (or companies). These products have some value to the purchaser. Marketing is a social process. The customer survey is the ideal way of gauging and understanding one's customer base.

The problem many companies face is listening to "their own hype" and not paying attention to the customer. Technologists are especially good at "drinking their own bath water." Many technologists tend to dismiss any comments unless they are words of agreement or praise. Once the company stops listening to the customer, the company is finished. Many Internet companies have gone out of business for several reasons. One reason is that they often provide a service or product that no longer has any value to the customer. Unfortunately, many of these Internet companies fail to understand the circumstances of their failures. Some blame the stock market and others even blame themselves. Before any business blames someone else for its failure, it must first look at whether or not they met the customers' needs. Figure 3-9 is a representation of how important customer surveys are to a business.

Focus groups are like customer surveys in that they obtain customer reactions and opinions. However, conducting focus groups is expensive compared to sending out a survey. Some marketing companies specialize in focus group meetings.

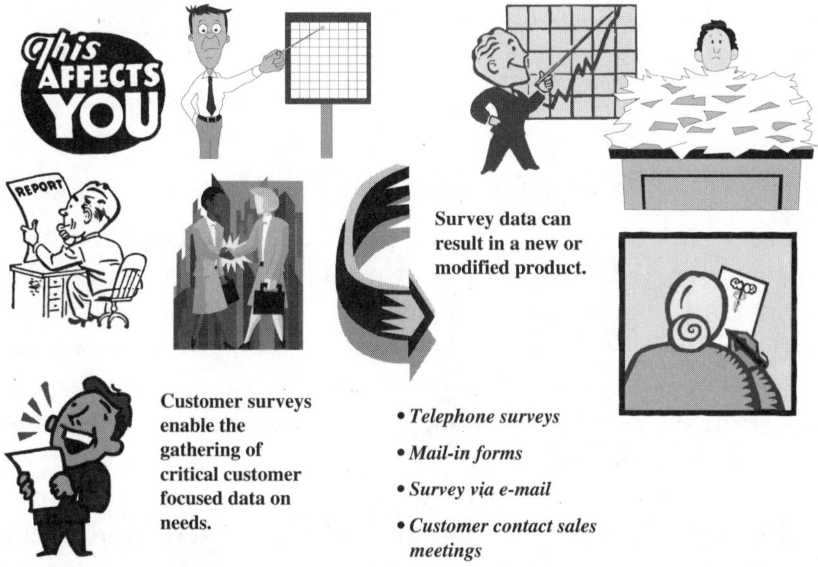

This AFFECTS YOU
REPORT
Survey data can result in a new or modified product.
Customer surveys enable the gathering of critical customer focused data on needs.
• Telephone surveys
• Mail-in forms
• Survey via e-mail
• Customer contact sales meetings

FIGURE 3-9 Customer surveys

However, the task of finding people to survey is a daunting one. Furthermore, the people who are taking part in a focus group study must be paid.

The following is a secondary set of tools that an ISP or e-commerce company can use for measuring customer satisfaction:

Complaint and Suggestion Systems Such a system not only enables a company to handle customer complaints, but also gathers information regarding a customer's likes and dislikes.

Analysis of Lost Customers Companies that lose customers ought to survey those customers that have decided to no longer do business with the company. Many lost customers would jump at the chance to let the disobedient ISP or other company have a piece of their mind. The complaints should be viewed as valuable customer input.

A customer in the Internet world can be examined, viewed, perceived, and understood in the same manner as one would in the non-Internet world.

DEFINING CUSTOMER VALUE AND SATISFACTION

In order to define the value of a product to a customer and the customer's satisfaction with that product, one needs to understand consumer behavior. Consumer behavior is a complex study of how people select, buy, sell, use, and dispose of products, services, or ideas in order to satisfy wants and needs. In one form or another, everyone is a consumer to some other party. Unlike the brick and mortar stores, the B2C Internet stores do not have face-to-face interactions with their customers. The lack of personal and physical interaction means that the B2C Internet store cannot obtain real time customer opinions. The question now is, how does the Internet-based retailer or commerce company obtain customer reaction? The answer is via customer surveys and real people conducting focus groups with customers. These tools are not perfect and very dependent on how the questions are asked. The Internet world operates by the same rules as the non-Internet world.

Consumer behavior can be characterized in a structured model. The consumer behavior model is known as the "seven O's":

Occupants	Who constitutes the market?
Objects	What product(s) is the market purchasing?
Objectives	Why does the marketplace purchase the product(s)?
Organizations	Who participates in the purchasing of the product(s)?
Operations	How does the market go about purchasing the product(s)?
Occasions	When does the market purchase?
Outlets	Where does the market conduct its purchasing?

Figure 3-10 illustrates the framework for modeling consumer behavior. The reader should note that in the world of e-commerce, the exact same factors in modeling consumer behavior appear as in a brick and mortar store.

The previous framework of questions requires an Internet company or a non-Internet company to understand their customers' backgrounds, personalities, and culture. The lack of a physical interaction with the customer presents a truly difficult obstacle to overcome. People's real time reactions are often times a more accurate gauge than a well-thought out response. Emotion is an important way of measuring customer satisfaction. When people are given an opportunity to "clean up" their language, the marketing research company loses the instantaneous reaction that is a true measure of customer reaction.

The Internet company must identify in detail those factors that influence the behavior of the consumer:

- **Cultural** Cultural factors have the deepest and strongest influence on the behavior of any consumer. Race, religion, region of birth, the values one grew up with, occupation, education, caste (social status), the socio-economic status

7 O's for Modeling Human Behavior

- **Occupants**
- **Objects**
- **Objectives**
- **Organizations**
- **Operations**
- **Occasions**
- **Outlets**

Important to understand the customers' backgrounds, personalities, and cultures.

Real time reactions are more useful to

FIGURE 3-10 Framework for modeling consumer behavior

in which one grew up, and the current socio-economic status of the individual are all cultural factors that influence behavior. The list is very broad and examines a person in total.

- **Social** Social factors include reference groups, family, roles, and statuses. Reference groups include groups that influence a person's behavior such as unions, professional societies, school clubs, social groups like the Knights of Columbus or the Masons, or even church groups. In general, families are considered the most influential group in one's life and also the most influential in buying disposable/perishable (food) and non-disposable products (homes, clothes, vehicles, and so on). Families encompass all of the aforementioned factors and even the following factors. They are a complete reflection of the individual buyer. Roles and statuses include those parts/roles one must play in public and in private.
- **Personal** Personal factors include the consumer's age, occupation, economic circumstances, personal hobbies, lifestyle, and self image. The reader may wonder what "lifestyle" means here. The word "lifestyle" is nebulous to many people. Lifestyle refers to a pattern of living. Specifically, it refers to how a person expresses themselves via their activities, interests (for example, likes and dislikes, what they read, and so on), hobbies, and opinions. Lifestyle represents how a person interacts with their environment.
- **Psychological** Four psychological factors influence a consumer's buying decisions: motivation, perception, learning, and beliefs and attitudes.
    - *Motivation* is a need that has reached a sufficient level of intensity in a person's mind, which has caused the person to take action to satisfy the need.
    - *Perception* is reality. Perception is a process by which a person makes decisions, organizes, and interprets information in a way that has meaning to the person.

- *Learning* is a process of personal growth whereby an individual learns via experience. One's experiences may include formal education or simply the process of living, working, interacting with people, and performing activities. People may buy a product or perform an action based on what they learn. Sometimes people learn when they take an action.
- *Beliefs and attitudes* are acquired when people learn or act. A belief is a thought that represents a person's opinion. An attitude is a person's evaluation of an event, and a lasting feeling towards an object or idea.

- **Buying Behavior** Buying behavior refers to the different types of purchasing based on the involvement of the buyer and the degree of brand differentiation. Four types of buying behavior exist.
  - *Complex Buying Behavior* occurs when a consumer is highly knowledgeable of a product and its competing brands and makes a purchase based on this body of knowledge.
  - *Dissonance-Reducing Buying Behavior* refers to the situation where a consumer makes a purchase and then discovers problems with the purchase. The consumer becomes so aware of the problems with the product that they begin seeking information justifying their decision. Usually, this occurs when the product purchased is very expensive.
  - *Habitual Buying Behavior* occurs when consumers often buy specific product brands out of habit.
  - *Variety Seeking Buying Behavior* is demonstrated by consumers that frequently switch brands for the sake of change, boredom, feeling of the moment or for any kind of reason.

Figure 3-11 is an illustration of the factors that influence buying behavior. Note that the illustration depicts how interrelated these factors are to one another.

Factors Influencing Buying Behavior

- **Cultural**
- **Social**
- **Personal**
- **Psychological – motivation, perception, learning, and beliefs & attitudes**

FIGURE 3-11 Factors influencing buying behavior

The Internet business environment has become an active and crowded space. Thousands of companies are seeking a multitude of opportunities in the Internet business. Companies are entering the business as middleware providers, end user application providers, terminal device manufacturers (especially the wireless device vendors), banner advertising companies, and ISPs. Each one of these companies fills a need or role in the overall picture of building a product or providing service. Today, Internet companies have sought to create a variety of relationships that will enable them to jump-start and expand their businesses. The idea of creating business relationships that are synergistic and enhance one's company is not a new idea. The maverick nature of the Internet entrepreneur and the newness of the Internet industry had led many early investors to believe that basic business practices are not applicable. In reality, what has been observed in 1999 is that basic business practices (such as finding allies) and rules of finance are not only applicable, but also undeniable realities of any business. In the year 2000, Internet companies are working feverishly to bring products and services to market by creating alliances. The Internet space is so crowded that many companies have been forced to develop a value chain analysis in order to determine how to best identify their place in the Internet business model and potential partner types. This examination should result in the creation of a value chain analysis.

THE VALUE CHAIN

A value chain is a business tool that is used to describe the value of a sequence of events or relationships. The value chain concept can be applied internally and externally. From an internal perspective, a company would look within itself and identify a set of specific activities/events that create value and cost in its own company. These internal activities/events are classified into two categories. Each of the categories can be further subdivided into activities/events:

Primary Activities/Events

Inbound Logistics Material or information that has been shipped into the company.

Operations Departments that convert the material or information into some other form either for internal processing or for sale to the customer.

Outbound Logistics The shipping of product, whether a physical product or information, to the customer.

Marketing and Sales An external activity that results in the sale of the product to the paying customer.

Service The servicing of the customer. All products, physical or informational, need some kind of servicing to the customer. Service can include repair and maintenance, customer complaint desks, 24 hours seven days a week network support, or even free monthly software updates.

SECONDARY ACTIVITIES/EVENTS

Procurement Procurement is the purchasing of products, material, information, and so on. for the support of each of the primary activities.

Technology Development Technology development is usually seen as hardware or software. I tend to look at this from a broader perspective. An individual should look at the primary business of the company and he/she will find that a kind of structured development always can occur to improve the product. For example, a cookie manufacturer is always looking at how to make a tastier cookie. The process of making this cookie is always under review and requires a structured process. Development is a process of creation that ultimately results in a product. Making a cookie is food science. Making furniture is carpentry, which is a skill that has to be translated from a manual operation to a mass produced function. In summary, technology development should be viewed as something more than electronics and software. In the world of the Internet, it is software.

Human Resource Management *Human resource* (HR) management is an often overlooked and under-rated activity. In fact, HR management is an important activity. HR is the department that finds all of the skilled labor, which ultimately produces the company's product. An Internet company requires highly skilled labor. Finding highly skilled labor requires a good HR department.

Infrastructure Any company's infrastructure includes general management, finance, strategic planning, accounting, public relations, and government and legal affairs. Infrastructure activities are a result of all the primary activities and other support activities.

A company's success depends on how well each department performs its respective function and how well each department coordinates with one another. The ultimate goal for each department is to do their jobs well in order to deliver, in a timely fashion, a product that a customer wants to spend money on. Figure 3-12 is a depiction of the value chain model.

The challenge the Internet companies face is the pace at which the business is moving. The pace is so fast that Internet companies are working to create strategic alliances with other companies at a pace the general business community has not experienced since the days of the space race (circa 1959–1969). The value chain model can be applied to external relationships, that is, strategic alliances. In the case of the external application of the value chain model, one would overlay the model of

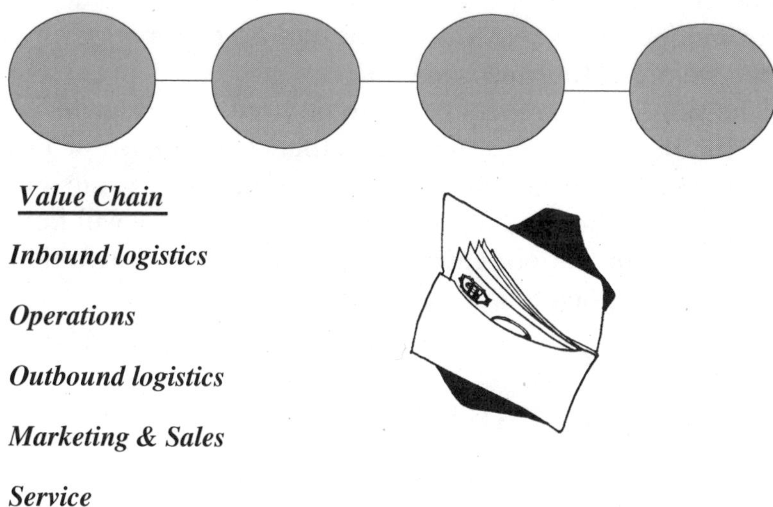

Value Chain
Inbound logistics
Operations
Outbound logistics
Marketing & Sales
Service

FIGURE 3-12 Generic value chain model

one company with the model of the other company. An intercompany value chain model would focus on the product and the areas of internal expertise of each company. Focusing on internal expertise enables companies to find ways of helping one another bring their products to market and possibly create new products. In the case of the strategic alliance, one company wants to have the role of the primary customer contact.

The Internet is filled with a variety of players all wanting to get customers to buy from them. Today, the large telecom carriers are working feverishly to become the ISP of choice. Internet hardware and software companies want these carriers to buy their software platforms, information databases, and hardware. However, this flurry of activity has caused the large carriers to say, "I want to deal with one company that has the relationships I want to help me be successful. I am tired of speaking to hundreds of companies that are all providing the same thing and claim to be better than the other." Today, the Internet software and hardware business has created the role of integrator. The integrator is a company that serves as a primary point of contact and service and hardware integration for the carrier to communicate with. It becomes incumbent upon the integrator to ensure that their partners are the "best of breed." The value chain model analysis identifies the points of synergy for the respective allied Internet companies.

The value chain model can also be used to overlay revenue flows for the companies allied together. In the case of provisioning to the telecommunications carrier, the value chain model can also be used to explain how the revenue will flow to the telecommunications carrier. Figure 3-13 is an illustration of the external application of the value chain model.

INTERNET MARKETING

As the reader may have noted by now, I do not see marketing the Internet any differently than marketing any other product or services. Marketing is a process by which people obtain what

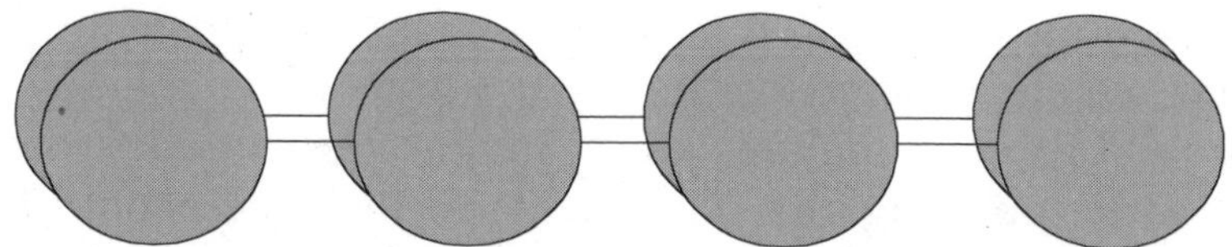

A Value Chain analysis can be performed of the value two or more companies bring to one another in a relationship.
Overlay revenue flows to analyze the revenue relationship between two or more companies.
Value Chain
Inbound logistics
Operations
Outbound logistics
Marketing & Sales
Service

FIGURE 3-13 External application of the value chain model

they need and desire through creating, offering, and exchanging products with others. These products have some value to the purchaser. Marketing is a social process, that involves communicating a message to the masses of people. The sales process is about closing "the deal" with a specific customer. Salespeople use marketers to identify target market segments.

Marketing to the Internet customer base is no different than marketing in any other industry. It is about fulfilling needs, wants, and demands. As cliché as this statement appears, it is the truth. Salespeople merely close the deal. However, the sales effort would fail if the marketing effort was not at least adequate. Identifying customers, market segments, demographics, demand levels, profitably levels, and future needs are all part of the marketing effort. Salespeople focus on their own needs–specifically, converting the product into hard cash. Salespeople assume that customers will not buy unless they are convinced, sold on, and promoted to about a product. Marketing people take an outside-in look at their own company, where they are concerned about satisfying the needs of the customer base first, followed by examining (and executing) all the activities associated with satisfying the customer's needs. Salespeople look from inside their company and then outside to the specific customer. They are focused on selling because their salaries are normally commission-based. Any reasonable person selling would not care about the overall needs of the market. Figure 3-14 illustrates the marketing concept.

Marketing and sales are two diametrically opposed efforts that cannot exist without one another.

RETAIL MARKETS

Marketing across the Internet is functionally no different than marketing any other product. The Internet is a medium, so the marketer promoting the product needs to determine how to use the medium to their advantage. Companies selling products must understand that any company needs an image or message

Marketing people want to satisfy the needs of the customer first and then worry about how their company will meet their needs.

Marketing people are focused on understanding a market, not a specific customer.

FIGURE 3-14 Marketing concept

to communicate to its customer base. The Internet retailer must create and maintain an image. Some Internet retailers are known for selling books. These same retailers have opened up their site to the sale of other kinds of products. Suddenly, the nation is joking that this Internet retailer is selling barbecue grills. The result just described might have been part of that retailer's market plan. However, it is doubtful that this Internet retailer was seeking to become the target of jokes. Image is everything in marketing.

The retail chain Sears, Roebuck & Co. is known for items such as vehicular tires, hardware, and home appliances. This chain of stores is very successful in its market space. Internet sites that retail products to the consumer are no different than brick and mortar stores. The customer must associate himself/herself with the Internet retailer with a specific product type, otherwise the retailer becomes everything to everyone. The reality is that the shopping patterns of people are affected by the image of the store. Some people would never be caught in a discount store known for selling clothes that are not manufactured by high-priced fashion houses. Discount brick and mortar stores even segment their market space. Some discount stores focus on appliances whereas others focus on popcorn, junk food, and clothing. Internet retail sites are no different. Internet retailers should know the customer segment to which they are communicating. Figure 3-15 depicts the need to establish and communicate an image.

During the 1999-2000 holiday season, a large national toy store started selling products over the Internet. To many people, this event was a confusing one. The toy store promoted itself as a place where any toy a person wanted could be purchased. The toy store even promoted fast moving checkout lines. The toy store was a national chain, discount store that had a presence in nearly every community in the United States. However, the toy store's Internet site was not adding any additional discounts. The question everyone was asking was why bother selling over the Internet? This resulted in confusion on Wall Street. The final result was total disaster for the toy store chain. Customers' orders were either filled late or unfilled. Customers became

Images are as fragile as glass.

FIGURE 3-15 Image

angry and the toy store's stock (Wall Street) took a nose dive. Selling over the Internet does not mean instant success. Planning and execution are still important.

PLANNING AND EXECUTION

The planning process described earlier in this chapter portrays the overall planning process. Planning is an activity that can be divided into strategic and tactical categories.

Strategic planning The process of developing and maintaining a viable fit between the company's objectives and resources and the changing marketplace. Strategic planning looks at the future of both the company and the marketplace.

Tactical planning The process of developing and maintaining a viable fit between the company's objectives and resources and the marketplace at the moment. Tactical planning is a management process that supports the strategic plan of the company. Simply put; tactical planning is putting the company's strategic plan into action—here and now. It is a short-term activity.

Execution is the process of implementation. The execution of plans requires a day-to-day management of resources in order to fulfill company objectives. The best-laid plans will fail if the execution is not flawless. Execution requires identifying resources, assigning resources, meeting established timelines, and care. Execution is comprised of four basic elements that must exist in harmony, otherwise the chance of a successful implementation is lessened:

Common way of thinking and behaving Refers to style and corporate culture.

Staff Refers to the type of talent the company has employed to execute the plan.

Skills Refers to the specific training the staff has undergone and the skill sets they currently possess.

Shared values Means that all of the employees share the same values and work ethics.

When all four elements are present and in harmony, one will find companies that are likely to succeed. Investors who have seen Internet companies fail can usually recognize the absence of these elements as a reason for failure. A brief note: Many of the ISPs that failed in 1999 and 2000 failed because of poor management and leadership. Poor management and leadership have a direct and deadly effect on the four elements of execution. Figure 3-16 depicts the elements of execution.

Just Do It!!
• Common way of thinking
• Talent
• Skills
• Shared corporate values

FIGURE 3-16 Elements of execution

MEASURING MARKET DEMAND

Market demand means the total volume of product that would be bought by a defined customer in a defined geographical area over a defined period of time in a defined marketing environment. Forecasting tools, sales figures, sales projections, and surveys are used to measure the market demand. The Internet employs site hits as a way of measuring customer demand or interest in its site.

The type of Internet site will determine the importance of the site hit. The site hit is merely the user entering the Internet company's Web site. This does not mean that the user has purchased anything. It just means that he/she has seen the primary Web page of the Internet site. The site hit is similar to a television entertainment program's rating. The rating determines the popularity of a program by counting the number of people in a defined pool of people watching the program. Using statistical analysis, a national viewing percentage is calculated. Internet companies that are nothing more than search engines use the hit as a way of determining their popularity. This number can then be used to charge companies fees for advertising space. The more hits, the higher the advertising fees. A television news program that also has a Web site can use the hits to help determine which kinds of stories to give greater weight. The hit is the easiest method of determining a level of consumer interest. Figure 3-17 represents how market demand is measured in the Internet business space. Market research companies are already employing more advanced methods to predict demand more accurately. These methods include internal processes review, purchased external research, observational research, focus-group research, and a number of quantitative tools such as discriminant analysis and sales-response modeling.

THE BUSINESS MODELS

The business models to date have kluged together efforts. Some Internet companies have failed and others have thrived. The successful models have taken on the form of existing business

Many Web sites measure market demand based on the number of hits, number of resultant purchases, and a group of other variables.

FIGURE 3-17 Market demand

models in the non-Internet world. These models have taken on new names such as those in the following list:

- B2C
- B2B
- X2X
- B2G

The Internet models are different than non-Internet business models in that the Internet reaches a larger audience and does not have the benefit of real time personal interaction. The revenue layer of the model is no different than any other revenue model. People and companies perform a service or sell a product and expect to get paid. One of the important things to understand is that the value chain model must clearly articulate

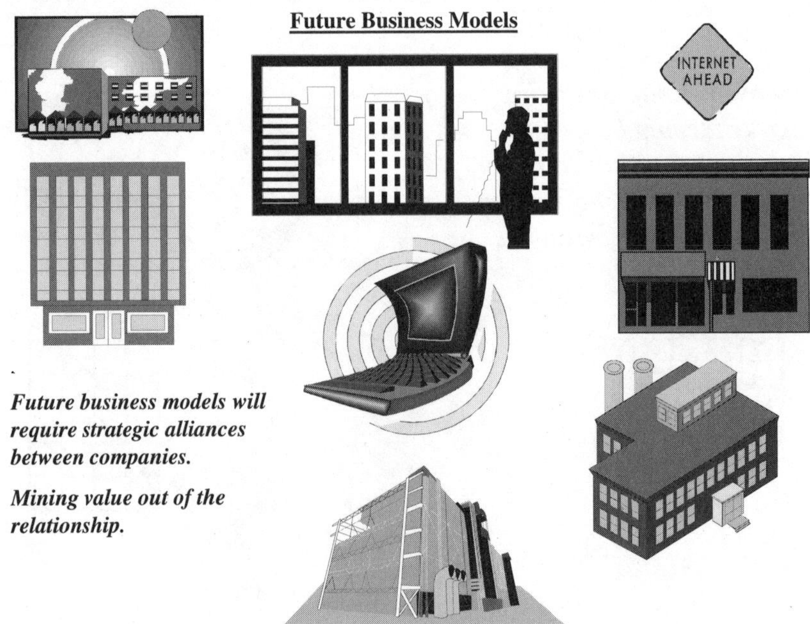

Future Business Models
INTERNET AHEAD
Future business models will require strategic alliances between companies.
Mining value out of the relationship.

FIGURE 3-18 The future business model for the Internet

the parties involved in bringing the product to market. Furthermore, in the Internet business environment, the elements of value are often software and not hardware.

The business models of the future will depend on a framework in which people can clearly see the various chains of revenue. The application of existing business principles will serve as a foundation for all future Internet business models. Figure 3-18 is a depiction of the future Internet business model.

TECHNICAL CHALLENGES

Much of the interest in the Internet is on the business side. The technological aspect of the Internet will be given attention later in this book. The reader should note that to many people, the technology is much further along than the business. For many years, especially the last few years, the technology community spoke voluminously about the benefits of the Internet. The reality is that the business side has spent the late 1990s and the year 2000 catching up. The entire nation is undergoing a cultural transformation simultaneously; the Internet has infiltrated the business, home, and now the schools. However, the technology community is facing challenges.

These challenges are only becoming more defined in 2000. As the business opportunities become more apparent, so will the technical challenges. To the layman, the technical hurdles are as follows:

More bandwidth to the home People want more than just 57,600 bits per second over the telephone lines. More bandwidth translates into video and voice applications. In the year 1999, video and voice applications became increasingly popular. The customer base is expressing interest in voice and video to the extent that the incumbent local exchange carriers (telephone companies) have launched an all out effort to enter the Internet business.

Faster computers Today, the desktop computers operate at speeds up to 700 MHz. People want more speed.

More memory The more memory in a computer, the more complex the applications one can place on their home and office computer.

Bigger hard drives The need for storage has increased to the point where companies are selling external hard drives and external disk systems capable of storing hundreds of megabytes of data.

The reader should not think the previous customer perceived technical challenges as trivial. In reality, the aforementioned list impacts the buying decisions of the consumer. The buying decisions ultimately have an effect on their ability to use the Internet. Perception is reality in marketing. Consumers will perceive an ISP as having problems if the consumer's interaction with the ISP appears slow or flawed, even if it is due to a slow computer or a computer that does not have the software to view programs because of missing computer applications. Customer frustration has a way of impacting all parties in a service provisioning chain. See Figure 3-19 for a depiction of the customer satisfaction model from the technical perspective.

To the technology community, the challenges are exactly the same. However, their list is longer. In addition to the aforemen-

Customers want:
More bandwidth
Faster computers
More memory
Bigger hard drives
Customers perceive technology and its challenges differently than the technical expert.

FIGURE 3-19 Technical challenges—Customer perception

tioned, the following can be added to the list of technological challenges faced by the Internet community:

Network Security The current Internet protocol, known as IPv4, does not contain security features that ensure private transmissions. Note the message one receives from their ISP when they access a Web site that is not under the control of the ISP. The current protocol does not enable the ISP or the other Web site provider from wrapping the transmission with encryption algorithms. Every network, whether it is the wireline or wireless voice network, is vulnerable to security breaches. The new network protocol, known as IPv6, is far more secure than IPv4. IPv6 is theoretically capable of providing an environment conducive for creating defenses against active attacks, passive attacks, and replay attacks.

The *Encapsulating Security Payload* (ESP) is a fundamental element that provides integrity, authentication, and confidentiality for IP datagrams. The ESP works between hosts, between a host and a security gateway, or between security gateways. The support for security gateways permits trustworthy networks behind a security gateway to omit encryption while using security gateways to obtain confidentiality for transmissions over untrustworthy network segments.

Service Level Performance Service level performance objectives or quality of service objectives are a commonplace measurement of performance in the wireline telephone community. The various public utility commissions measure the telephone companies against a pre-defined set of performance parameters. If the telephone company fails to meet those objectives, they are fined or penalized by the utility commission in some manner. These performance parameters keep the telephone companies focused on serving the customer. ISPs do not have that culture of service ingrained in them. IPv4 is not robust enough to maintain a voice conversation without information bits arriving at the incorrect times.

The problem with packet networks is that the static and warbling are an annoyance, but a reality, in a network type whose primary asset is in its ability to packetize voice and transmit the packets across different routes. One can buffer non-voice data, but one cannot buffer voice. Buffering voice results in delayed speech, static, and to some people, a warbling type of sound.

Scalability One could say that the current Internet is highly scalable. However, for those ISPs attempting to enter the voice business, the issue of scalability becomes a challenge. Voice traffic has characteristics that are different than non-voice data. It is true that the average Internet call lasts approximately two hours and that the average wireline telephone call lasts for approximately 10 minutes. One would assume that the Internet is far more robust given the length of time a user will stay on the Internet. However, a network undergoes far more usage stress when it must respond to multiple requests for service. Compare the robustness of a single network switch capable of handling 400,000 calls in one hour versus a single network host capable of handling 40,000 users in one hour.

Network Management Network management in the world of the Internet has no central point of network management. Internet providers manage their own networks, but do not communicate with one another at the network management level. In the wireline telephony environment, network management centers are monitoring traffic between each other. Network operation centers in wireline telephone companies communicate congestion situations to one another. Neither wireline telephone company type (local and long-distance) wishes to be blamed for poor service; therefore, it is incumbent upon them to ensure good service.

Sometimes network congestion reaches a point where a wireline telephone user hears the words "all circuits busy." This does not help either the local or long-distance telephone companies. Customers encountering delays tend to get mad at both carrier types. Add video and the challenge becomes even greater.

Network Reliability No ubiquitous network reliability standards are employed in the Internet community. Network reliability standards exist in the traditional telecommunications carrier world. Reliability is defined as availability, throughput, and latency (time delay). *The Federal Communications Commission* (FCC) oversees a national forum known as the Network Reliability Council, where carriers and infrastructure vendors participate in the establishment and maintenance of network reliability standards. The FCC performs this function in order to ensure that the various networks maintain a common standard of reliability for the purposes of national security.

The focus of the Network Reliability Council is the health of the physical infrastructure and the associated practices. ISPs have no forum and do not participate in the FCC's Network Reliability Council. ISPs have an additional concern that the wireline and wireless carriers do not have currently; ISPs must be capable of supporting third party applications. ISPs cannot easily control the impact of third party applications on reliability. Today, most of the ISP solutions are proprietary. ISPs wishing to enter the larger market of voice provisioning must be capable of meeting the standards of reliability that the other telecommunications carriers are required to meet.

It should not be assumed that any of the challenges listed are "show stoppers" for an ISP wishing to expand his/her business beyond the e-mail business. Technology usage in the Internet is more than the use of applets, Web page creation software, search engines, audio programs, video viewing programs, streaming audio and video programs, or any other desktop software. Technology in the Internet also means network connectivity and communication. The biggest technical problem for the Internet industry is bandwidth. The bandwidth shortages affect every Internet player, whether they are an ISP,

- **Network Security**
- **Service Level Performance**
- **Scalability**
- **Network Management**
- **Network Reliability**

FIGURE 3-20 Technical challenges of the Internet community

ASP, e-commerce player, or an m-commerce player. The inability to move vast amounts of data quickly is the Internet's biggest problem and is the one that is currently under the control of other carriers—namely the large, wireline telephone carriers.

Figure 3-20 illustrates the technical challenges faced by the Internet community. The technological challenges of the Internet will be further explored in Chapters 4, "Internet Technology and Service Provisioning in the Information Marketplace" and Chapter 5, "Network Signaling: Mobile and Wired Business Space."

SUMMARY

The Internet is a new telecommunications medium that has just begun to be understood, explored, and/or exploited by the technical or business communities. The need to expand the Internet business will require the business community to find ways of finding new market opportunities, improving current businesses and ways of doing business. The technology community must find ways to meet the needs of the marketplace. As of 2000, the Internet is still in its infancy. What we do understand is that every sector of the telecommunications business is attempting to be in the Internet space.

CHAPTER 4

INTERNET—TECHNOLOGY AND SERVICE PROVISIONING IN THE INFORMATION MARKETPLACE

The Internet players, specifically the ISPs (Internet Service Providers), are currently working toward entering the voice business space. The ISPs are facing an uphill battle. A major issue that an Internet player faces when entering the voice space, which is occupied by the existing wireline and wireless carriers, is lack of understanding of the voice business. The voice business is not a primitive technology or business space. It represents the mass market. E-commerce and m-commerce notwithstanding, the mass market is a market of voice users. E-commerce and m-commerce mine enormous additional value out of the network. However, before the ISPs reach that point, they are attempting to provide the most basic need—voice.

Voice is the most basic and sought after service across the entire telecommunications marketplace. Despite the various data tools and services available, people need to communicate verbally. Internet companies understand that voice is the most important means of communication, which is why the ISP is seeking ways of provisioning voice. Once the ISP has the user, he/she can take full advantage of providing e-commerce and m-commerce opportunities. The Internet players must understand the technology basics of their competitors' networks: the existing wireline and wireless carriers. However, the important point to understand is that network planning in both the Internet and non-Internet world is fundamentally the same.

This chapter will focus on helping the reader understand the wireline and wireless carriers' basic network technologies and practices. A startup ISP needs to understand that voice is more than just *Voice over IP* (VoIP). Many non-technical and technical issues should be considered when providing voice. The primary purpose of this chapter is to assist the Internet professional in understanding the importance of voice. Understanding voice will equip an ISP with the fundamentals of providing any service to the user.

The role and impact of the *Application Service Provider* (ASP) will be described later in the chapter.

NETWORK PLANNING

Planning a network involves understanding the technical needs of the service provider's network and the financial realities of fulfilling the technical requirements. ISPs face the same problems.

To the wireline and wireless carrier, the single largest component of the network tends to be transmission facility-related (trunks, signaling links, and so on). Wireless carriers are wireless only between the radio base station and the handset. Communication connectivity between the base station and switch is made by physical and in some cases, air-link transmission technologies. Planning the physical network involves understanding how to achieve a balance between network optimization and cost control. Network planning is an exercise in achieving a balance between technical functionality (facility and switch) and cost control.

Network planning is essential to any service provider, whether or not they are an ISP. When planning a network, whether it is to support Internet service, satellite service, or paging service, a set of common planning (design) activities are available. A description of the planning stages should give the reader a better understanding of just how common or synergistic many networks are.

Some basic tools are needed to design a voice network. These tools are applicable to all network types. They are not software or hardware, but are conceptual in nature:

- Architectural switching plan—This refers to the efficient routing of calls or data packets based on network efficiency or business arrangements. Route diversity is a major concern for many service providers. But this requires a dollar cost.
- Transmission plan—Types of facilities used and transmission design requirements.
- Addressing plan, numbering plan, point codes, and the like.
- Network signaling plan (more detail in a later chapter).
- Subscriber services plan (more detail in a later chapter).
- Subscriber billing/customer support plan.
- Network management and operational support systems plan.

This book will not address all of these tools in detail. The intent of listing these planning tools is for informational purposes only. More detail on this subject can be found in my book *Telecommunications Internetworking*.

ARCHITECTURAL SWITCHING PLAN

Until 2000, most ISPs were only concerned about the following architectural issues:

- Geographic placement of the servers in a local market—The location of the ISP's point of presence determined the telephone numbers the ISP was able to obtain.

- The *local exchange carrier's* (LEC's) local switch had enough capacity to handle the traffic to the ISPs servers.

The aforementioned list of issues explains why so many individuals and "mom-n-pop" shops (local drug store, shoe store, candy store, and so on) were able to enter the ISP business space. At one time, anyone could install a computer and modem, and receive a telephone number from their local telephone company in order to become an ISP. It was and still is that simple to become an ISP providing e-mail and Web search capabilities.

The largest e-mail company today faces staggering administrative overhead in order to manage services and servers in every telephone area code. ISPs want their customers to dial into their systems without having to incur long-distance charges. The creation of the mega-carrier, where local and long-distance telephone companies and wireless carriers have merged, has changed the carrier landscape. ISPs will be charged long-distance access fees by the local telephone company because the Telecommunications Act of 1996 and state laws enable the local telephone company to recover its cost of providing the ISPs network access to the long-distance companies. However, the mega-carriers wishing to provide Internet service will not have those same fee issues.

If ISPs are hoping to become more than just e-mail companies, they will need to engineer and install network systems that provide more control over their costs and ability to create services and deliver services. The existing wireline and wireless carriers understand how important it is to have control over their own customer accounts and access to the customer. This is why the wireline and wireless carriers own their switches and transmission networks.

Owning one's own network is an expensive option compared to leasing a transmission network. However, to the customer, owning the transmission network enables the service provider to more accurately control overall network costs. To the wireline and wireless carriers, paying a leasing fee for the use of a transmission facility means a loss of control over how to optimize the network for performance. Leasing has its benefits, but if one had to lease their entire transmission network, his/her costs would be astronomical. The overall cost of being an ISP is enormous because all of the ISPs are pouring money into another service provider's pockets for local and long-distance access. Installing your own network represents a high initial capital cost, but it also represents long-term savings in outside fees and surcharges. In 1999, the United States government allowed the LECs to charge the ISPs access fees for long-distance connectivity. Achieving a balance between network performance and cost control is a complex, but essential task for any provider of services. This exercise is called *network planning*.

The wireline and wireless carriers use three basic network configurations to move voice traffic: Tree, Ring, and Star. The configurations were designed for non-packet, hierarchical routing schemes. Hierarchical routing is a must in non-packet services. In a packet data world, these configurations are applicable to both the local loops and the regional routing configurations that an ISP would be concerned about. These configurations can also be used to support services, such as voice, that require a fixed routing time period. In a packetized world, disaster recovery is an old concept. Alternate routes are always a necessity. ISPs rely on nationwide network providers that have routing diversity designed into their networks.

The nationwide network providers are not Internet or packet network providers, but are long-distance carriers selling network time to another service provider. These long-distance networks are still point-to-point networks. It is true that these carriers are working to install packet technologies. However, $200 billion of embedded capital is in the network today, and it is not going to be that simple to replace in total. The following sections represent how these nationwide network providers make nationwide connectivity for the Internet.

TREE ARCHITECTURE The Tree architecture looks like a tree. The characteristics of the Tree architecture are the same as those of a typical *local area network* (LAN), some landline telephone company networks (party lines), or even cellular carrier networks. The Tree architecture is the most efficient way of distributing the same set of communication signals to multiple terminals. Figure 4-1 illustrates the Tree architecture.

RING ARCHITECTURE The Ring architecture loops traffic (voice or data) so that it returns to its original starting point. Some examples of the Ring architecture are the Token Ring and the Self-healing Ring. Service providers (wireline and wireless carriers) most commonly use the Self-healing Ring.

The Self-healing Ring is a ring architectural configuration in which the same messages are transmitted simultaneously in opposite directions on parallel rings. In case of a service interruption on the ring, the messages are automatically transferred to the other ring. This provides a high degree of redundancy. At each port, node, or drop on the ring, messages can be trans-

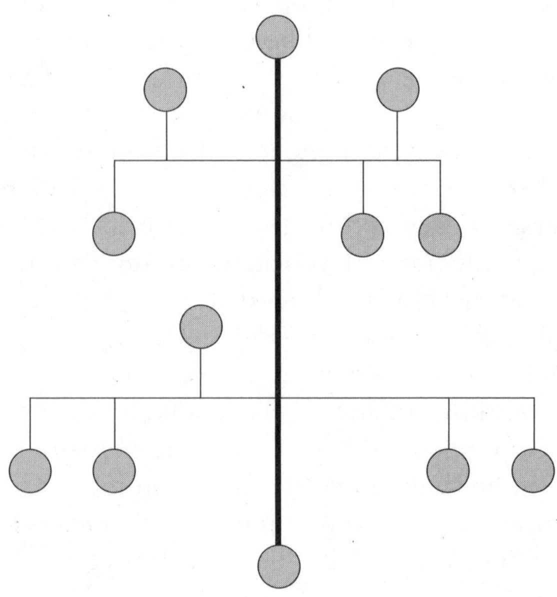

FIGURE 4-1 Tree architecture

ferred to a Tree, Star, or another Ring network. The Self-healing Ring is a network design used to ensure that the network continues to provide service in the event of a network failure. This design methodology/philosophy is called *disaster recovery planning*. The ISP must understand that no network is failure proof. Therefore, steps must be taken to ensure continued service. The concept of redundancy in the world of computer systems is not a new concept. However, to a service provider, disaster recovery capability means more money and higher operating costs.

Ring networks for cable television, landline telephone companies, or wireless carriers normally utilize fiber optic facilities for digital transmission. The Self-healing Ring is expensive because one must support at least two networks that are carrying the exact same messaging for the sake of redundancy. However, the expense is worth the disaster recovery capabilities inherent within this architecture.

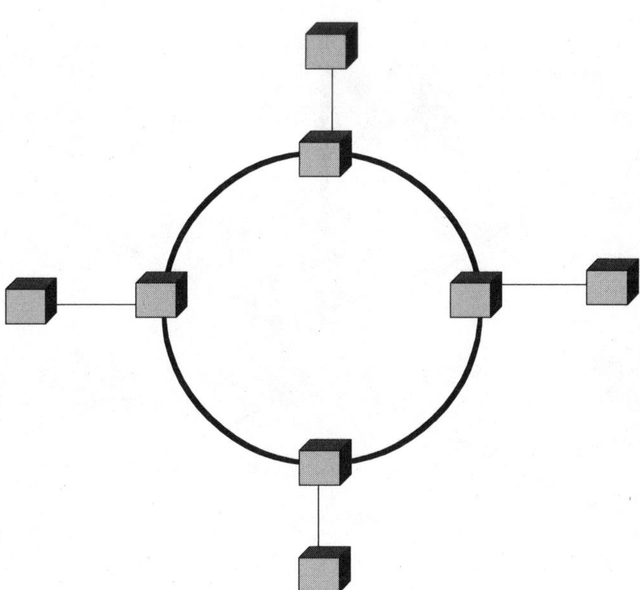

FIGURE 4-2 Ring architecture

STAR ARCHITECTURE The Star architecture is the most typical wireline telephone company configuration. The Star topology enables separate transmission paths to be established to each subscriber. Each path can be designed to carry the same or different messages.

The Star architecture is the most efficient way for the network switch to communicate to a large number of end users. The Star network configuration can be likened to an airport hub. In this case, the hub would be a telecommunications hub.

ROUTING TECHNIQUES

Although the Tree and Star architectures are common in the cable television and telephone industry, the Ring architecture is being increasingly used. The need to recover from network failures and damage is important and is becoming increasingly difficult to meet as the networks grow in complexity. The architecture of a network plays an integral role in maintaining

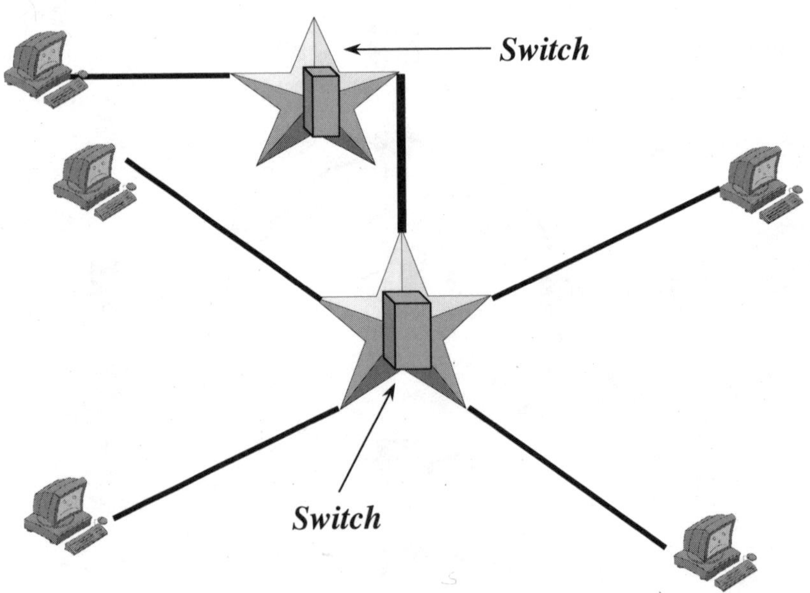

FIGURE 4-3 Star architecture

network health. Many people may believe that the data in packet networks are safe from loss during network failures. That is not true; all networks can fail. Packet networks use a routing technique known as dynamic routing. However, dynamic routing does not necessarily mean that the data packets will reach their destination. Physical transmission routes fail. Disaster recovery is a network planning methodology in which networks are designed with "imminent failure" in mind. Disaster recovery requires one to design redundancy in his/her network.

The questions of how much redundancy and how much network protection to design into a network are difficult to answer. A crossover between network cost and network disaster recovery occurs. The point at which a carrier or Internet carrier reaches too much network protection depends on network design policies and corporate financial objectives. In other words, it is left up to the carrier to decide what is enough or not enough.

It is a useful mental device to think of networks like vehicular roadways. Networks have figurative signs instructing drivers when to Stop, Yield, make a Left Turn, make a Right Turn, Merge, Go, and Slow Down. Figure 4-4 illustrates this mental device.

FIGURE 4-4 Networks: Vehicular roadways

Routing is the action of directing information from one point to another point. Three principle forms of routing are used in the traditional telecommunications world:

- Hierarchical routing (also known principally today as Alternate routing)
- Dynamic routing
- Packet routing

However, by taking a more expanded view of telecommunications and including the data world, a third type of routing exists. This form of routing is called packet routing (that is, packet switching of information).

Routing is not a transport technology. It is a methodology of moving information around the transport structure, which will be discussed in this chapter. The network configuration is interrelated to the routing methodology that can be optimally executed. Given its importance, the next section will elaborate further on this topic.

The original form of routing first used in the old, wireline voice telecomm environment was hierarchical routing also known today as alternate routing. Alternate routing is applied to voice traffic by providing a first choice (high usage trunk route) and one or more alternate routes if the first route is unavailable. This form of routing is hierarchical in nature because the switching system follows a predetermined order of routing. Dynamic routing lends itself to flexibility and grew out of the need for more efficient routing and management of the enormous volumes of traffic crossing the nation. Dynamic routing is similar to packet routing in the area of flexible management. Packet routing entails the routing of information where each packet takes a different route to the final destination.

HIERARCHICAL/ALTERNATE ROUTING

A pure Internet configuration does not lend itself to the structured and planned routing methodology of hierarchical/alternate routing. This form of routing has its advantages and disadvantages. However, it is important for all Internet companies to understand that the embedded wireline network is interconnected to all ISPs and providers of other Internet services and therefore, becomes a potential "chokepoint" for high-speed services. Figure 4-5 is an illustration of a hierarchical network.

DYNAMIC ROUTING

Dynamic routing refers to the updating of routing patterns on a real time or near real time interval on the basis of traffic statistics collected by a *network management system* (NMS). Dynamic routing is a traffic routing method in which one or more central controllers determine near real time routes for a switching network based on the state of network congestion measured as trunk group bust/idle status and switch congestion. The choice of traffic routes is not predetermined.

Dynamic routing schema are used in situations where the complexity of routing the call is so high that applying the fixed rules and structure of hierarchical network configurations make

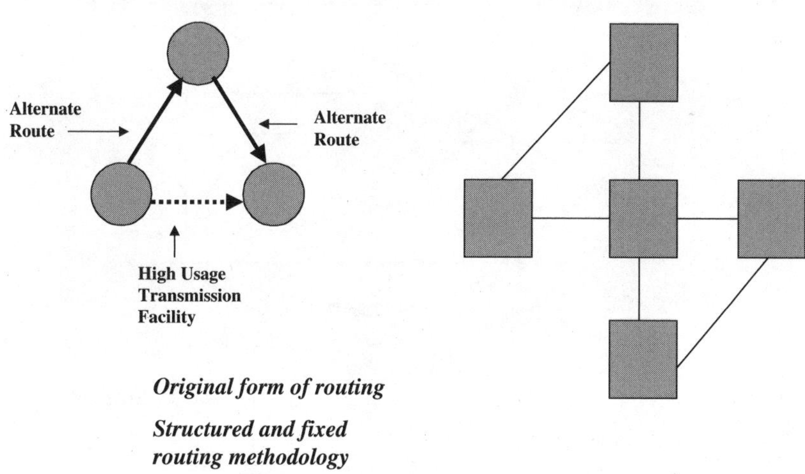

FIGURE 4-5 Hierarchical/Alternate routing

it difficult for a network manager to properly manage call flow. To some extent, a degree of intelligence must be embedded within the network.

The following diagrams compare the complexity of routing information in a hierarchical routed network versus routing information in a dynamic routed network. Figure 4-6 is an illustration of the dynamic nature of the dynamic routing schema.

Figure 4-7 exhibits an even more complex set of network relationships. It is a pictorial description of the Internet, exhibiting a number of networks. These networks are interconnected in a variety of ways. Given the changing regulatory environment, the networks will be interconnected in this seemingly random fashion. Business relationships dictate how the networks of various service providers are interconnected. Therefore, even though the networks appear to be interconnected in a seemingly random fashion, they are connected in a realistic fashion. Applying the rules of hierarchical routing is nearly impossible in the Figure 4-6, yet it is this routing methodology that enables today's ISPs to route across the country.

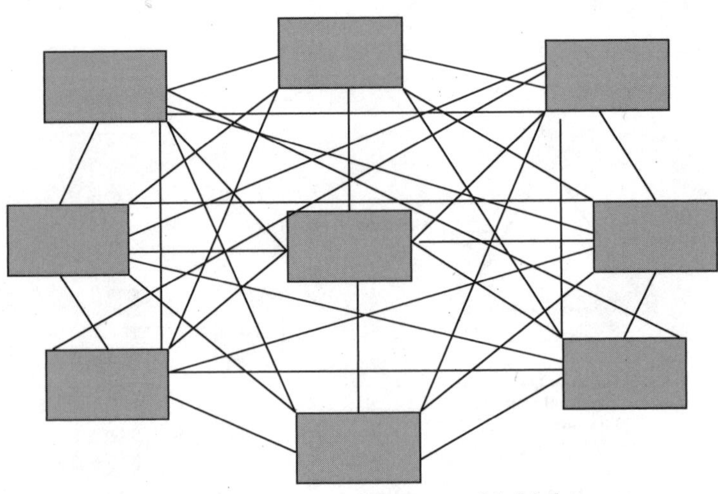

FIGURE 4-6 Dynamic routing

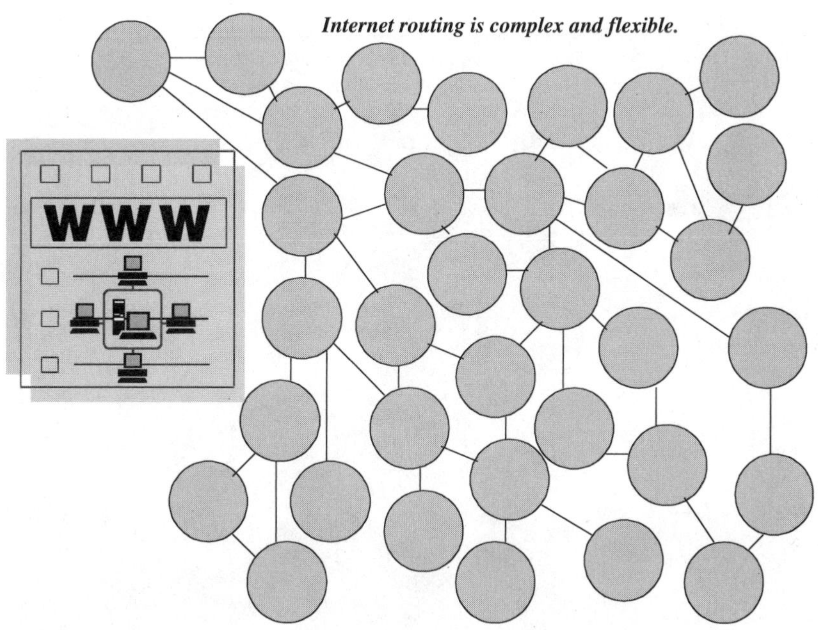
Internet routing is complex and flexible.
WWW

FIGURE 4-7 The Internet: Complex network routing

PACKET ROUTING

Packet switching represents one step beyond traditional voice telecom switching and is suitable for a data environment. Packet switching involves the transport of information in discrete packets or packages of information. Each packet is tagged for identification with its label, origination, and destination. The packet is launched into a network conditioned to support packets of information (rather than a continuous stream of information), and ultimately, it finds its way to the destination of choice. Each packet takes a different route to the final destination. This is not exactly a dynamic routing schema, but fairly close to one because the routes are not predetermined and change constantly. Figure 4-8 depicts how a stream of data packets will traverse the Internet from its point of origin to its destination.

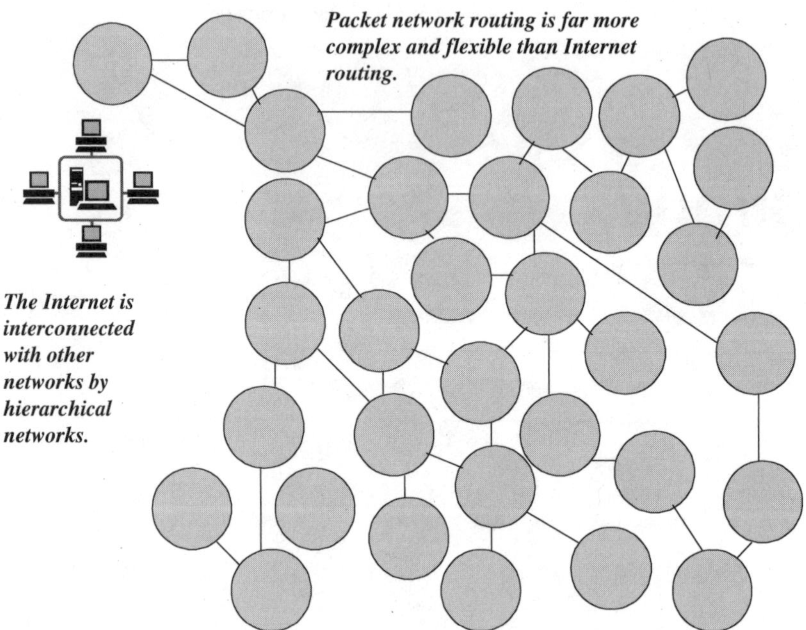

Packet network routing is far more complex and flexible than Internet routing.
The Internet is interconnected with other networks by hierarchical networks.

FIGURE 4-8 Packet data networking

The wireline Internet is conceptually supposed to be a packet network. In reality, hierarchical networks interconnect the Internet with other networks. The packets stream across the nation via a fixed route across the wireline telephone network. When the packets reach the local market, in other words the ISPs' servers, the data stream may traverse the local market via a packet network. At this time, the financial realities show that the wireline telephone companies are the dominant network providers, the dominant telecommunications carrier presence in all local markets, and control the local loop. Due to the wireline telephone company's dominant presence, the ISP reaches the home and small businesses via a twisted pair of local loop wire. Figure 4-9 is an illustration of how the ISP is interconnected to the overall larger and dominant wireline telephone network.

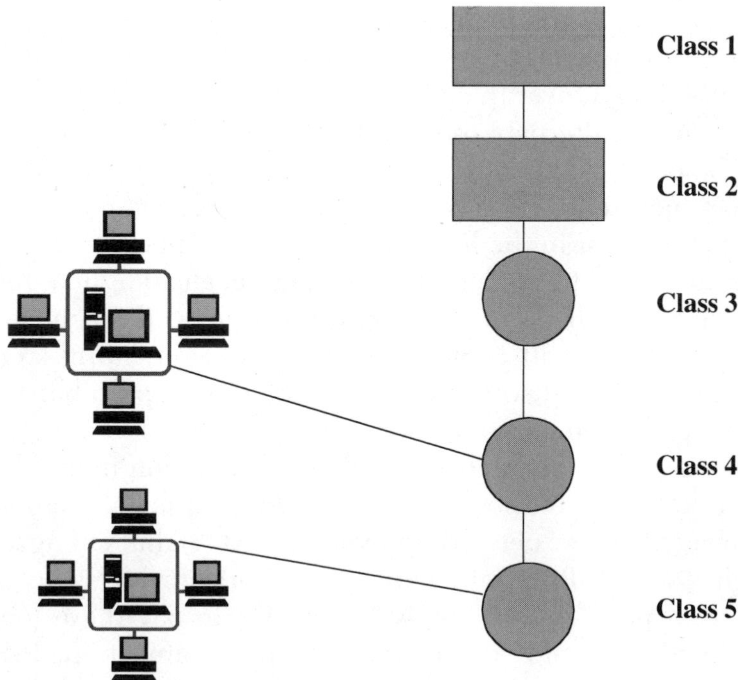

FIGURE 4-9 The Internet and network interconnection

The topic of network interconnection is a separate topic that requires a separate book alone. Further information on network interconnection can be found in my book *Telecommunications Internetworking*.

NETWORK ROUTING AND FINANCIAL RELATIONSHIPS

A relationship exists among the number of trunks, the financial cost of the direct route and alternate route, and the total dollar cost for serving the given offered load. The high usage facility cost is proportional to the number of high usage trunks. If no high usage trunks are available, all of the traffic must be sent on the alternate route, making the incremental alternate route money cost high. If a high usage group is available, the incremental cost of the alternate route is lower because theoretically, less traffic is sent over the alternate route.

The cost of alternate routing decreases the minute the first trunk is added to the high usage trunk group. The flip side of this cost picture is that a point is drawn where too many high usage trunks exist in the high usage group. As more high usage trunks are added to the high usage group, each additional high usage trunk will theoretically carry less traffic, while each alternate route trunk will continue to carry a significantly larger amount of traffic. At some point, it becomes cost prohibitive to add any more high usage trunks.

Network costs are a component of any decision in building a network. No matter how much route diversity is built, a point is drawn where a network provider must claim enough is enough. Today, ISPs require national network providers in order to go from point A to point B. ISPs need to pay these long-distance providers for the privilege of interconnection. ISPs' cost structures can blossom beyond all expectations resulting in bankruptcies.

The various routing schema described is applicable to networks of different service providers. A set of rules that defines how the flow of information is managed between networks must be present. This does not refer to signaling protocol like TCP/IP, but rather the management process of sending information between networks.

Regardless of the regulated or non-regulated nature of any information service provider, a methodology of how information is sent from point A to point B must be enforced. Today, some ISPs own the servers in a local area, but do not own the transmission media connecting the servers. In fact, some companies transport information across the nation yet do not have visibility to the consumer (the individual). The methodology behind the flow of information affects how the information will be transported and who it will be transported by, which ultimately impacts the cost of the network. Figure 4-10 illustrates the need to understand the financial aspects of planning a network.

Other Costs

- **Servers**
- **Software**
- **Data Archiving**

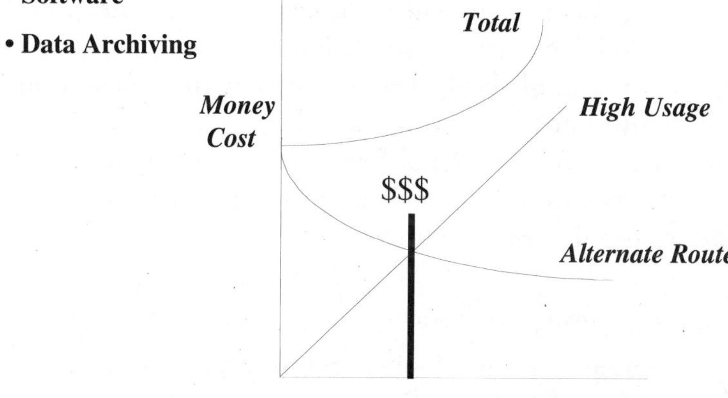

FIGURE 4-10 Network costs

TRANSMISSION PLAN

One should always keep the following in mind in relation to transmission facilities:

- Should I lease or buy?
- What is the initial capital outlay for owning the transmission facility?
- What are the maintenance costs?
- What are the recurring fees (if you do not own the transmission facility)?
- How do I get to point B from point A with minimal financial exposure?

One cannot overstress the importance of financial awareness in the managing of any business. Technology and ideas are important to a business, but money pays the bills.

In the case of non-voice telecommunications networks (that is, data or the Internet), a pattern appears to the way network information flows. Currently, ISPs need to connect to a LEC in order to transport their call/information from one point to a point outside of, or even inside the LEC's jurisdiction. Even in the current world of the Internet, the flow of information from an ISP to another ISP maintains order.

Once the routing schema has been determined, the following questions should be asked about the network and the number of connections required:

- What are these connections?
- What transport technologies and protocols are available to support the network configuration?
- What type of transmission facilities should be used?

The aforementioned questions comprise the transmission plan. When a call is made, data and video are transmitted, and whether it is wireless or landline-based, facilities are used for transmitting the call. These facilities are also called transmission facilities.

Transmission facilities can be broken down into the following basic types:

- Metallic
- Fiber optics (glass)
- Microwave

The major difference among facility types is speed (sometimes referred to as bandwidth). Speed refers to the speed at which information is transmitted. Time is money and the higher the information speed supported by the facility, the more money is made and saved.

Transmission facilities' sole function is to serve as the distribution media for information. However, the bulk of the transmission facilities in use today throughout the plethora of networks is a limited bandwidth local loop wire, DS-0 (supports up to 57,600 bits per second), or a T-1 (supports up to 1.544 mega bits per second).

VoIP is possible over the limited bandwidth transmission facilities because of buffering. Despite the speed of the computer and the various network elements comprising an ISP's system, the transmission facilities used by the telephone companies in the local loop do not enable high quality (perceived by the listener) voice. A few missing bits of information will cause the listener to hear static. What hurts VoIP as a service currently is that the packets of voice data will arrive out of sequence and at different times, resulting in delayed speech. Missing packets of voice data makes it worse for the listener. Even digital transmission does not mean that a data network does not suffer lost bits. All data network providers follow network performance standards. Selecting the right transmission medium is an important step that is often overlooked. Today, most ISPs are grateful for connectivity to the user, which is no small feat given the fact that the facility being used is either leased or resold from the telephone company. The type of medium dictates the speed or in other words, the bandwidth capacity that is possible.

ADDRESSING/NUMBERING PLANS

A numbering plan or destination address describes the location of the originator and receiver of the call.

PUBLIC TELECOM SERVICE PROVIDERS: THE TELEPHONE COMPANIES AND WIRELESS CARRIERS The first numbering plan originated in the wireline telephony world. A numbering plan is simply the telephone number of the subscriber. In North America, the telephone number is a 10-digit number. The format of this 10-digit number is:

- Numbering Plan Area (NPA) + Number Exchange (NXX) + four digit station number (XXXX)
- The NPA is the three-digit area code and its format is NXX.
  - N is any digit two through nine.
  - X is any digit zero through nine.
- The Number Exchange is a three-digit code that is typically associated with the service provider's switch. Its format is NXX.
  - N is any digit two through nine.
  - X is any digit zero through nine.

In North America, the NPA defines a specific geographical area. Given the growing shortage of numbers, a new numbering format has been adopted called the Interchangeable NPAs. The Interchangeable NPA format involves interchanging an NPA and an NXX. In other words, in the past, the NPA-NXX combination 301-555 would not be re-used as 555-301 because it violated past numbering format rules, which stated that the first digit could be any number from two through nine; the second digit could only be a zero or a one; the third digit could be any number from two through nine. Those rules made sense at the time. However, given the massive growth of telecommunications services and products requiring telephone numbers, a new format needed to be adopted (the Interchangeable NPAs). These services range from office fax machines, home fax machines, new service providers requiring numbers, and so on.

As of 2000, the wireline and wireless carrier environments are running out of 10-digit telephone numbers. The FCC is currently administering the process through industry selected third companies. Carriers once were able to hold onto telephone number blocks for years without ever using them, but the FCC has stopped this practice. The list of measures being taken by the FCC in cooperation with the carrier community is extensive. At some point, the industry will have to expand the telephone number beyond 10 digits.

INTERNET ADDRESSING The Internet uses an addressing scheme that appears to the layman as a fairly flexible identification plan. A subscriber can select a name or number that has meaning only to the subscriber or one that identifies the subscriber. The only limitation is that someone else cannot be using the Internet address already. The Internet is a whole new type of network that does not seem to fit the traditional telecommunications business or network model. However, the Internet is the new major player on the block. This will be discussed in more detail later in the book.

The Internet uses TCP/IP for transmission. TCP/IP is a simple data protocol. The mapping of mobile station addresses to an Internet server can be approached in the same manner as approaching protocol conversion, for example, X.25 to SS7 or GSM MAP to IS-41 MAP. The Internet address is composed of a source IP address, source TCP port, destination IP address, and destination TCP port. The new Internet protocol, IPv6, expands the addressing scheme even further. More on this subject will appear later in the book.

As ISPs take a larger role in the provisioning of services to the user, at some point, all of the public obligations that the wireline telephone companies and the wireless carriers are held accountable for will also apply to the wireless carriers. These obligations include 9-1-1, law enforcement surveillance, and a quality of service standard.

NETWORK SIGNALING PLAN

A service provider communicates with other service providers and network elements in its own network using a specific type(s) of network signaling. Network signaling refers to the language (protocol) the network "speaks" to other networks and within its own framework. Network signaling plans identify the signaling protocols that will be supported by the service provider.

Several different network signaling protocols are in use today. Each signaling protocol is optimized to function in specific environments for specific reasons. The following represents the more popular protocols in use today:

- *Signaling System* 7 (SS7)
- *Asynchronous Transfer Mode* (ATM)
- Frame relay
- TCP/IP
- *Multi-Frequency* (MF)

More information on these protocols will appear in the next chapter. Today, industry work is examining how to overlay the Internet protocol over some of the aforementioned protocols. The wireless industry is currently working toward implementing *Wireless Access Protocol* (WAP) and i-mode. Figure 4-11 represents the various protocols in use today.

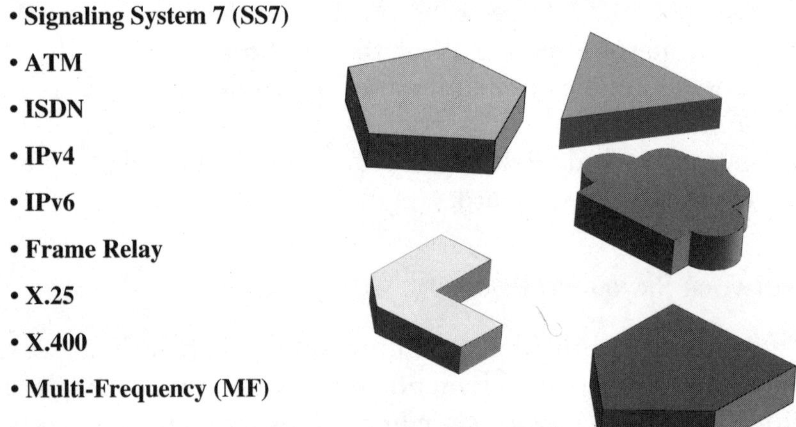

• Signaling System 7 (SS7)
• ATM
• ISDN
• IPv4
• IPv6
• Frame Relay
• X.25
• X.400
• Multi-Frequency (MF)

The network of networks utilizes a variety of network signaling protocols.

FIGURE 4-11 Network protocols

NETWORK MANAGEMENT PLAN

The network management plan refers to the systems used to manage the service provider's network. The NMSs would include the following:

- Network element management
- Network systems management
- Service management

Planning of these systems requires coordinating functions between each of the systems listed. If the plan is executed properly, the result will be a well-run *Network Operations Center* (NOC). In general, the network operations manager is concerned with information and system tools that enable him/her to run/manage the network. Managing the network involves scheduling activities and resources, managing network traffic load, routine maintenance, trouble reporting, and diagnostics. A generic description of network management would be "network maintenance and health."

All of the following are applicable in the managing of any network, including an ISP network.

NETWORK ELEMENT MANAGEMENT As noted in Chapter 1, "Networks and the Internet: What Are They?", network element management requires the ability to monitor and manage specific network elements. A network element can be defined as a database, the switch, a router, a voice mail recording system, an adjunct system, or an application service provider.

- **Maintenance schedules** Maintenance schedules refer to the scheduling of activities concerned with the provisioning of high quality subscriber service and network stability.

- **Routing table update schedules** Routing table update schedules refer to the scheduling of updates to the switching system's routing tables. The term "routing table" has its historical roots in the traditional public telecom world.
- **Number of subscribers being served by the system** Knowing the number of subscribers should be a given. Without knowing how many customers are being served, the network manager cannot properly manage the traffic load of the network. Traffic load management entails moving telecommunications traffic from one part of the network to different parts of the network and even blocking new traffic to ensure that existing traffic is routed properly.
- **Services being provided to the subscribers** Network managers need a complete picture of the services being made available to the subscriber base. These managers are not concerned about who specifically has what subscriber service, but are concerned about the services available to the population of subscribers as a whole.
- **Performance objectives for the network element** Measuring the performance of a network element is way of measuring network health. The network manager must have an understanding of how each element in the network is expected to behave. Using these benchmarks, the network manager can address malfunctions or anomalous conditions.
- **Transaction processing time thresholds** Processing time of any transaction is another indicator of network health.
- **Types of diagnostic tools are available to him/her**
- **Operating procedures documentation**
- **Failure levels/indications** Network elements will fail at some point in the life of the device. The failure needs to be reported to the network manager.

- **Alarm conditions indications** As indicated earlier, failures and anomalous conditions need to be announced in some manner. The alarm indications should reflect via visual or audible means.

NETWORK SYSTEMS MANAGEMENT Network systems management involves the way one manages the entire collection of elements and all the functions within the business that affect the physical network. Network systems management is about the total managing of the network and its components.

Network operations managers seek to have the same questions and issues addressed about the whole network as they would the individual network element. These questions/issues are as follows:

- Service objectives for the system.
- Network monitoring software tools.
- Network diagnostic tools.
- Anticipated traffic load on the system.
- Overall traffic profile of the system on an hourly and daily basis.
- Types of subscribers being served.
- Level of network system control—can the entire system be re-booted or does is an involved element-by-element process involved?
- Can traffic be re-routed?
- Types of alarm indications.
- Types of network failure indications.
- Overall network operating procedures documentation.
- Any disaster recovery plans.

Managing a network is similar to taking care of one's own physical health. Some medical doctors specialize in specific areas of medicine and specific parts of the body. Other doctors (even today) are generalists; they look at the person as a whole.

Managing a network requires the network manager to be a generalist. The network manager should be able to understand each network component while simultaneously understanding how each network element relates to each other. He/she needs to assess the health of the network from a micro and a macro level.

SERVICE MANAGEMENT Service management links the network systems management function with the customer care function. This includes performance metrics and customer satisfaction metrics. Service management requires direct involvement with the customer. It involves every mechanism and process required for delivery of services to the customer, that is, operational support systems, sales, and marketing. Service management links the network to the financial/revenue portion of the telecommunications service business.

SUBSCRIBER BILLING/CUSTOMER SUPPORT PLAN

The customer has a role throughout all aspects of a service provider's business. The customer is at the heart of the carrier's product offerings, the managing of the network, and even the network architecture of the network. Yet most providers create plans that address subscriber billing/customer support. The customer is so vital to any business that the treatment of the customer is singled out as a major planning effort. This type of plan does not involve marketing to the customer exactly. More information on customer support is available later in this book.

The following diagram, Figure 4-12, illustrates the relationship of customer support with the overall business:

APPLICATION SERVICE PROVIDERS The *application service provider* (ASP) is a company that writes software that is used on a carrier's network. The ASP in the Internet world writes not

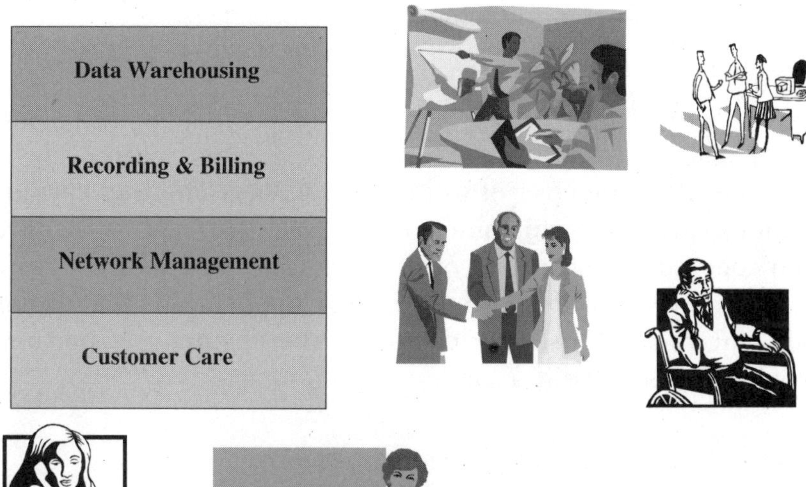

FIGURE 4-12 Customer support

only the enabling software but also provides, in some manner, content or some type of content-related service. ASPs today are serving as gateways to customer bases. Thousands of ASPs claim to have access to huge pools of customers either directly or through their own service partners. The services are provided using ASP-specific software tools. The ASP may provide a variety of services such as:

- Internet directory access
- Video and audio content
- Internet radio
- Polling
- Games

- Stock exchange monitoring
- Online purchasing
- Online stock trading

The ASP is the layer of value that the large Internet carriers cannot bring to the customer themselves. The ASP is focused on a specific service or set of services that enhances the need for the network. ASPs develop software applications that do the following for the customer or enable the customer to perform activities better or differently:

- Make certain activities easier to perform
- Information that is easier to find
- Faster to do
- Less expensive to do
- Quicker to travel to
- More interesting and higher in quality to view
- More interesting and higher in quality to listen to

The ASP will typically sign an agreement with a large Internet carrier. This agreement will often times state that for a price, the ASP will give the large carrier access to a pool of subscribers and use the ASP software to enable the service. This pool of subscribers has already expressed a desire to purchase a service from the ASP, but the ASP needs the carrier to physically access the customers. The service in question is usually an application that the ASP has written but will not license to the carrier for its use. To many ASPs, the agreements maintain their control and ownership of the customer. This belief is a fallacy, for in the end, companies that can find a better way of providing the application will always exist. Furthermore, the carrier is the one with brand name recognition, bills the customers, and has the physical network in place. The goal of the ASP is to dominate the market and use that market power to either buy out a competitor or outsell them in the marketplace.

The ASP has changed the way the ISPs and other Internet "wannabees" do business and manage their networks. To a large ISP or other carrier, the ASP is a huge database with customers. The need to understand how these ASPs impact the ISPs' network is a new facet of network planning. Rather than simple network connectivity, the Internet business requires an intense level of database activity that has never been seen before. The Internet business has added inter-company database interaction as a major component of network planning. Figure 4-13 is a depiction of the ASP in the current network.

Typically, large carriers do not have their own in-house software development areas. Therefore, it makes perfect sense to purchase services or partner with an ASP that specializes in some specific type of software development. As indicated earlier in this book, administering a customer base requires an enormous expenditure of resources. ASPs enable a carrier to add value without the administrative overhead.

QUALITY OF SERVICE *Quality of service* (QoS) is a perception. In the case of the telecommunications industry, which the Internet is a part of, perception means customer perception.

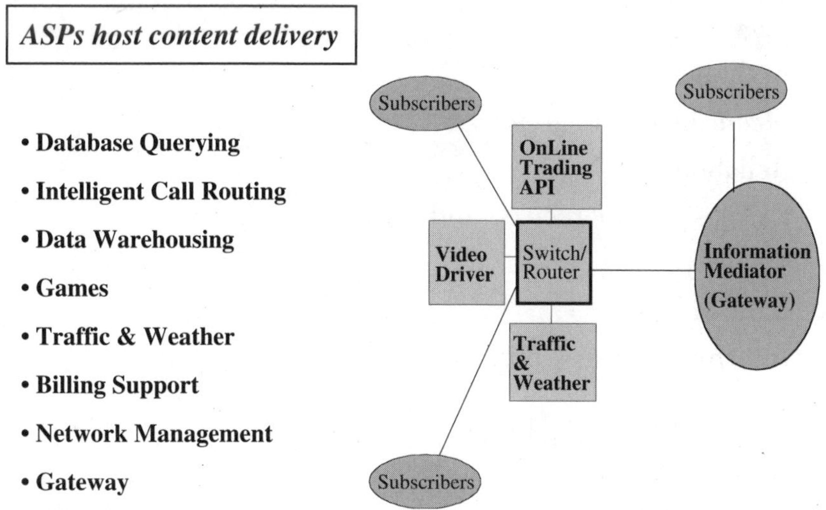

FIGURE 4-13 Application Service Providers

QoS is subjective; however, QoS can be measured using qualitative tools. From a customer perspective, QoS involves the following subjective needs:

- Does the service meet the customer's needs?
- Is the service easy to use?

QoS can be defined as the standard measurement used to determine whether or not the service provider is providing the service in a manner that meets the expectations of the customer. Despite the subjective nature of QoS, objective ways exist of establishing QoS levels. QoS is typically measured from the network standpoint. Measuring QoS from the network perspective is far easier than attempting to guarantee performance from a terminal device. The service provider does not normally provide terminal devices; therefore, the service provider should not be held to a standard of performance that involves terminal performance. Figure 4-14 illustrates how QoS is an objective and subjective measurement focused on the customer, yet based in the network.

QoS parameters can be divided into the following categories:

- Availability
- *Mean time between failure* (MTBF)
- Reliability
- Delay—both perceived and measured
- Security
- Bandwidth
- Information loss—bit error rate; video and audio

The aforementioned parameters measure specific things or levels of performance. Individually, the parameters mean nothing to a customer. However, when one views the parameters in total, the result is some overall perceived level of performance.

•Quality of Service (QoS) is a subjective measurement of system/service performance.

•QoS is a perception.

•QoS can be measured using qualitative tools.

•QoS is measure from a network perspective.

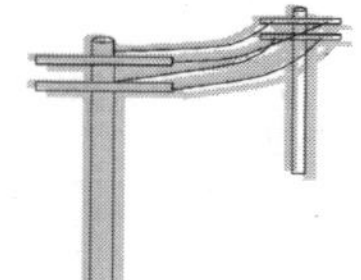

QoS is a mission critical standard to meet.

FIGURE 4-14 Quality of Service

QoS affects all players in the Internet business: ASPs, content providers, e-commerce and m-commerce providers, and so on. The reality is that users of e-commerce and m-commerce sites expect availability of a site 24 hours a day, 365 days per year. Figure 4-15 is a depiction of the QoS parameters. The following section is an explanation of these parameters.

AVAILABILITY Availability is the amount of time a system (computer, network, and so on.) is available for processing transactions. The way to measure availability is by taking the ratio of the total time a system is capable of being used during a given time period (the industry norm is one year). Therefore, availability of a Web site or some system providing an application would be measured as some number of hours over the total number of hours in a year. The result is typically presented as a percentage. The wishful objective of all system operators is 99.9999 percent availability. An availability of 99.9999 percent means that the system is only unavailable for 31.536 seconds per year. Most telecommunications companies can reasonably expect an availability of 99.97

Quality of Service Parameters

- **Availability**
- **MTBF**
- **Reliability**
- **Delay**
- **Security**
- **Bandwidth**
- **Information Loss**

The QoS parameters individually have no meaning.

QoS parameters are viewed in total for an overall understanding of performance.

FIGURE 4-15 QoS parameters

percent, which translates into a yearly unavailability of 2.628 hours.

MEAN TIME BETWEEN FAILURE (MTBF) ***Mean time between failure*** (MTBF) is the average time a manufacturer estimates a failure will occur in a piece of equipment. Individual components of a system have MTBFs that are different than other system components. Overall system MTBFs occur. These system MTBFs need to be qualified as total system outage. The time period given for a MTBF should be on the order of several months or years.

The MTBF will impact system overall system maintenance costs, which eventually impacts the cost of service to the customers andultimately, profit margins.

RELIABILITY Reliability is very different than either availability or MTBF. Reliability is a measure of performance or dependability, whereas the system is operational. It is a subjective measure that takes into account availability, MTBF, and consistent quality of product. Reliability is a subjective measurement because how much weight is placed over one measurement compared to another depends on the person.

TRANSACTION DELAY If a customer perceives that a particular Web transaction or even making a telephone call takes too long to complete, the company must determine what that perceived human limit is. Human factors engineering has always been a component of telecommunications network design. The human ear/customer perceives a delay within a range of 250 milliseconds to 500 milliseconds. After the user has entered the last digit, a telephone call may take as long as three full seconds to complete (the calling party hears ringing). However, a user who hears a delay in receiving dial tone longer than 500 milliseconds will perceive a problem with telephone service. A user is double clicking on icons to enter a Web site and it takes nearly a minute to actually see the next Web page. However, because the user has either experienced delays this long or even longer, he/she does not perceive a problem in this case. The delays are perception-based, but are indeed valid transaction delays that customers encounter.

Measured network delays can affect the QoS. These delays are in information transmission and packet data assemblage. These kinds of delays are information delay variations. Packet data is not an ideal way of supporting real time applications like voice conversations or video streaming. The delay variations must have defined limits set for each application so that no perceived degradation in QoS occurs. The delay variation limit may be 50 milliseconds. This would mean that the application being supported must have all of the data transported across the network within a 50-millisecond window. This window would require that the beginning and the end of a Web page reach the destination all within a 50-millisecond time period. The fact is that no industry standard governs the way in which Web pages are presented.

The initial transmission of voice over a packet network requires a sufficient number of voice samples to be collected before it is placed inside a data packet and then launched. The *Internet Protocol* (IP) data packet will place about 20 voice samples inside a packet. Voice is sampled at a rate of one every 1/8000 seconds in a digital wireline network. The math is very simple. A normal packet is approximately 600 bytes in length.

A voice sample is generated at a rate of one every 125 microseconds. Therefore, a packet delay would equate to 125 microseconds × 600 bytes = 75,000 microseconds. In reality, the IP packet can vary in length up to 64,000 bytes. When looking at the way packets are routed, one will recall that the packets are typically buffered at the destination end until a sufficient number of packets are assembled for final presentation. In other words, the delays can add up.

The problem with the Internet is that most people have come to expect delays. This is an unfortunate circumstance for it leaves many users with the perception that the Internet just cannot do it all. In reality, the industry has worked to meet the high consumer perceived standards of excellence by overcoming technical barriers and setting high operating standards.

Transaction delays can be due to limited bandwidth transmission facilities or inefficient and poorly designed Web sites.

SECURITY No telecommunications network is totally secure from fraud, computer viruses, and privacy threats. The wireline network used to live by the axiom: "The telephone network is inherently secure." This old saying was touted before the days of the computer virus, but even then it was untrue. Wire tapping and telephone bugging has been around for decades. The "bad guys" might not have been able to listen in on a phone conversation by tapping into a line while atop a telephone pole, but the "good guys" could tap into your phone lines from the switching center. Therefore, from a certain perspective, the telephone is inherently secure.

The information networks today are using a variety of methods to secure the network. These methods entail the following:

- Anti-viral software
- Passwords
- Encrypted transactions
- Network validation procedures

- User identification authentication procedures
- Fire walls

Security as a QoS parameter is an important one that is constantly being threatened by the hackers and criminals of the world. It is an unfortunate state of affairs that over a thousand new computer viruses are being discovered each month. These viruses threaten the personal security of the user as well as the security of the nation. A Web site that is not secure from unauthorized access invites trouble.

Security is even a bigger threat now to the Internet than it was a few years ago. The proliferation of ASPs has created a series of new opportunities for the hacker and criminal to wreak havoc on the network. Today, ASPs must pass a variety of requirements before an ISP will even consider connecting their network to the ASP's database.

E-commerce and m-commerce sites interact with users gathering data such as credit card information, home addresses, and home telephone numbers. Today, all information entered into most commerce sites is preceded with the statement: "You are entering a Web site that this provider cannot guarantee is secure from privacy threats." More users would exist if the e-commerce and m-commerce sites had a way of ensuring end-to-end security.

BANDWIDTH The more bandwidth available for transmission, the faster the download and upload speeds for data. The biggest complaint from Internet users is slow speed. The fact is that most ISPs have access to homes over a twisted pair of local loop telephone wire, which can only reliably support transmission speeds of up to 57,600 bits per second.

The lack of bandwidth has affected users' perceptions of a large number of Web sites that support video and audio applications. As a result of the lack of bandwidth, users cannot or will not bother accessing Web sites that provide high quality video and audio content. The larger the bandwidth, the greater the volume of data that can be transmitted per second. Voice and video require network access that supports real time applications. Network access refers to the modem, the local network, and the long-distance network. The lowest bandwidth device used in the network will limit the application.

INFORMATION LOSS Information loss is defined as missing bits within an information stream. This could be a large multimegabit size file that is missing some information due to noise over the transmission facility, some flaw at a network node, or even a node failure. Packet networks could suffer a network node failure, yet not lose all of the data because most of the data would be transiting the network via a different path. The network has no way of knowing that it contributed to a loss of information. The destination end would have to request the information file be resent.

Information loss parameters could be set where a transmission node or facility would have an operating standard with which to comply. The banking industry uses a standard of $1x10^{-9}$ bit error rate, which means that for every 1,000,000,000 bits of data, one can lose one bit of data. A standard of operations quality for a network element or facility should always be present.

The impact of data loss will vary from file to file and application to application. In a perfect world, one would not lose any data; however, this is not a perfect world and therefore, one will lose data despite operations standards. Transmitting the lost data would be possible in a text file. Some applications can recover from data loss and a retransmission request would enable a user to recover lost data. These applications could be a video or audio application; in such instances, the user would not be using these applications for real time two-way communications. The application itself cannot recover if the missing bits were sent; however, a retransmission request would enable the application to be resent in its entirety.

The potential for information loss increases with the number of parties now involved with the provisioning of an Internet-based service. Imagine the loss of information due to multiple network elements embedded within multiple Internet companies.

MOBILE ISSUES

All of the aforementioned sections contain information that is applicable to the mobile market. The mobile Internet faces a number of its own unique issues, which will be discussed in greater depth in the next chapter. Some of these challenges are as follows:

- Bandwidth
- Security
- Handset
- Reliability
- Development
- Time to market

These challenges are being addressed as the industry is rolling out services to the marketplace. The mobile Internet space faces challenges that the wireline Internet did not face.

SUMMARY

The marketplace for the Internet players is wide open. The ISPs are looking for subscribers. The e-commerce and m-commerce companies are looking to partner with ISPs. ISPs are planning to use voice as a way of grabbing customers. The e-commerce and m-commerce opportunities are a major portion, possibly even the majority, of the value the ISPs are seeking. However, without provisioning the most basic need, the ISP has no way of becoming that principal means of communication for the user.

The following chapter will examine the network signaling and transport technologies available to the Internet player. The kinds of players entering the Internet space will be discussed. Chapter 5, "Network Signaling: Mobile and Wired Business Space," will look at how current network signaling will support the Internet. Future signaling and Internet-specific protocol languages will also be discussed in this chapter.

CHAPTER 5

NETWORK SIGNALING: MOBILE AND WIRED BUSINESS SPACE

This chapter examines the network signaling available to the Internet player and will look at how current and future network signaling will support the Internet. Future signaling and Internet specific protocol languages will also be discussed.

One of the binding forces of network interconnection is the signaling. This network signaling is the language by which all service providers will communicate with other service providers and with network elements within their own network. A more technically precise definition of signaling is the exchange of information in a telecommunications network (public and private) that establishes and controls the connection of a call or a communication between subscribers or computing systems. It also establishes and controls the transfer of subscriber/end-to-end and management (subscriber and network) information. The signaling protocol of the Internet is the TCP/IP protocol suite.

Other networks supported by the wireline and wireless carriers support the following protocols:

- *Multi-Frequency* (MF).
- *Signaling System 7* (SS7).

- *Asynchronous Transfer Mode* (ATM).
- *Transmission Control Protocol/Internet Protocol* (TCP/IP).
- Frame Relay.
- *Integrated Services Digital Network* (ISDN).

The wireline and wireless carriers are seeking ways of deploying Internet capabilities in their networks. Some of their ideas involve the overlay of IP over their existing signaling infrastructures, whereas other ideas involve the creation of something entirely different. The wireless carriers have taken the bold step of creating new protocol suites to access the Web. Unfortunately, for wireline Internet users, the wireless Web protocol is not interoperable with the wireline Internet protocols.

The following section will provide an explanation of the underlying protocol suite, TCP/IP, in the wired Internet, and will explain the wireless protocol suite. Further details on the wireline network signaling protocols in the aforementioned paragraphs can be found in my book *Telecommunications Internetworking: Delivering Services across the Networks*.

TRANSMISSION CONTROL PROTOCOL/INTERNET PROTOCOL (TCP/IP)

The TCP/IP protocol suite was originally and currently is used for the internetworking of *Local Area Networks* (LANs). All of the signaling protocols used in the Internet are part of the TCP/IP protocol suite. The TCP/IP protocol suite was developed as a result of work first begun by the United States Department of Defense's *Advanced Research Project Agency* (ARPA) in 1957. The ARPA's objective was to develop science and technology in response to the military threat posed by the former Soviet Union. More information on the history of the Internet will appear later in this book.

TCP/IP protocol suite is composed of multiple protocols. The TCP/IP protocol suite is layered more than SS7 or even ATM. TCP and IP are just two of the protocols in the suite of Internet protocols. The term "TCP/IP" refers to this family of protocols. The TCP/IP protocol suite's multiple layers facilitate future development of new Internet protocols. Whether or not this was by design is irrelevant for this discussion; however, it is fortunate that the suite was constructed by the *Internet Engineering Task Force* (ITEF) in this manner because it has enabled software engineers across the globe to find new applications for the Internet.

The protocol architecture of TCP/IP was designed for use by the United States military. Given its roots, the protocol suite is capable of interconnecting multiple pieces of equipment from multiple vendors. The TCP/IP protocol suite has four layers:

- Physical/Link
- Network
- Transport
- Application

These layers correspond to the layers described in the *Open Systems Interconnection* (OSI) model. The OSI model was created by the *International Organization for Standardization* (ISO) for use in a computing environment. When communication is desired among computers from different manufacturers/vendors, the software development effort can be very difficult. Different vendors use different data formats and data exchange protocols that do not allow computers to communicate with one another. The OSI model is an engineering model that breaks everything down into simple and discrete tasks or layers. It serves as a framework for all telecommunications signaling protocol development.

THE OSI MODEL

The OSI model consists of seven layers. The communication functions are broken down into a hierarchical set of layers. Each layer performs a related subset of the functions required to communicate with another system. Each layer also relies on the next lower layer to perform more primitive functions and to conceal the details of those functions. It also provides services to the next higher layer. The layers are defined in such a manner so that changes in one layer do not require changes in other layers. By partitioning the communication functions into layers, the complexity of the protocol becomes manageable.

The following is a description of the layered architecture starting from the bottom of the stack.

Physical Concerned with transmission of an unstructured bit stream over the physical link. It invokes such parameters as signal voltage swing and bit duration. It deals with the mechanical, electrical, and procedural characteristics to establish, maintain, and deactivate the physical link.

Data Link Provides for the reliable transfer of data across the physical link. It sends blocks of data (frames) with the necessary synchronization, error control, flow control, and other overhead information.

Network Provides upper layers with independence from the data transmission and switching technologies used to connect systems. It is responsible for establishing, maintaining, and terminating connections.

Transport Provides reliable, transparent transfer of data between end points. It provides end-to-end error recovery and flow control.

Session Provides the control structure for communication between applications. It establishes, manages, and terminates connections (sessions) between cooperating applications.

Presentation Performs generally useful transformations on data to provide a standardized application interface and

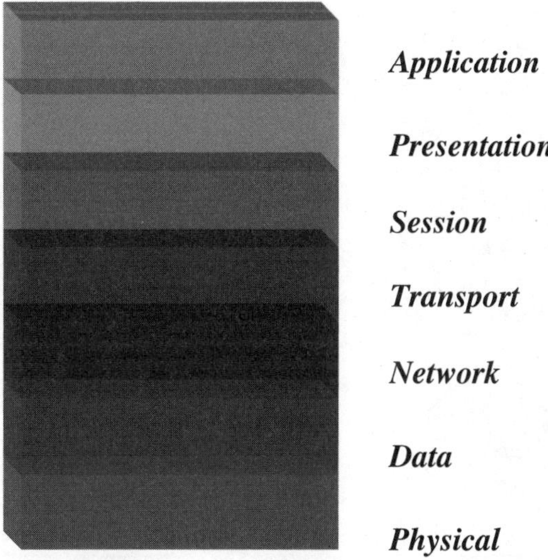

FIGURE 5-1 OSI Model

common communications services. It provides services such as encryption, text compression, and reformatting.

Application Provides services to the users of the OSI environment. It provides services for FTP, transaction server, network management, end users services, and so on. Figure 5-1 illustrates the OSI model. Notice that the following model represents a stack of layers. The foundation layer is the Physical Layer. Every other layer is built on top of the Physical Layer and each other. One can see how these layers are interdependent upon each other

TCP/IP MODEL

When one overlays the TCP/IP protocol model over the OSI model, direct correlations can be drawn between the two models. The TCP/IP protocol architecture is described as a four-layer model. Figure 5-2 is a comparison of the TCP/IP model with the OSI model. The layers of the TCP/IP model are described in the following list:

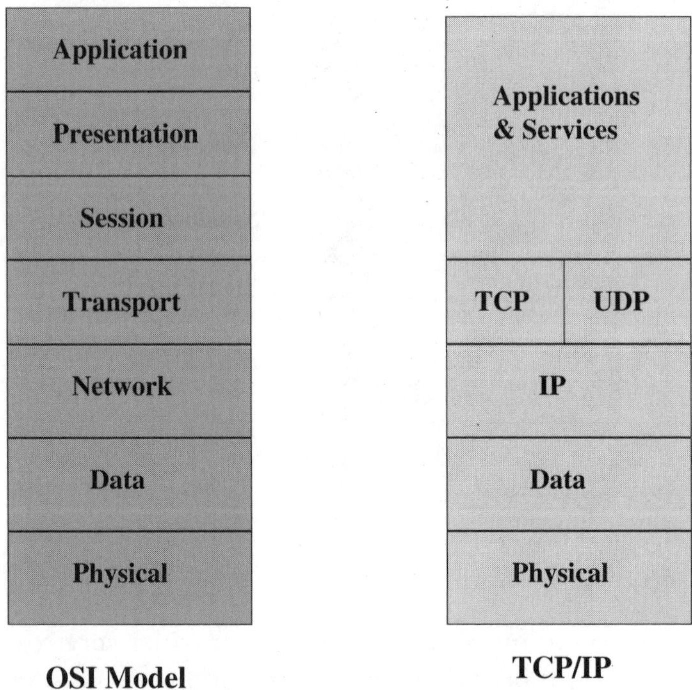

FIGURE 5-2 The OSI model compared with the TCP/IP model

Physical/Data Link Layer (as it corresponds to the OSI model's Physical Layer and Data Layer) also known as the Network Interface Layer, manages and routes the exchange of data between the network device and the network. The data or information referred to includes header information/overhead information.

Network Layer (as it corresponds to the OSI model) also known as the Internet Layer. This layer is responsible for managing the IP. The IP provides the Internet addressing for routing. The IP is a connectionless protocol that provides datagram service. A datagram is a method of transmitting information. The datagram is broken up into sections and is transmitted in packets across the network. More information on this layer will follow.

Transport Layer Corresponds to the same Transport Layer in the OSI model, transports the data. The TCP is run at the Transport Layer. More information on this layer will also follow.

Application Layer Responsible for managing all services. This layer corresponds to the Session, Presentation, and Application Layers of the OSI model.

The bulk of the capabilities behind the power of the Internet is found in the Network and Transport Layers. Therefore, additional detail is provided in the following subsections.

NETWORK LAYER

The Network Layer, also known as the Internet Layer or the IP Layer, supports connectionless datagram routing. Each datagram is routed along an independent path. Unfortunately, the IP does not guarantee delivery or even in-sequence delivery of the datagrams. Typically, a datagram is comprised of header information and the information packet itself. The header information is comprised of the packet's destination and source address.

In general, the problem with packet switching is that it has no way of guaranteeing the arrival of all the datagrams. Furthermore, packet switching does not even guarantee that the information delivered can even be assembled in the correct order. The IP is a packet switching protocol. Figure 5-3 is an illustration of a datagram.

TRANSPORT LAYER

The Transport Layer in the TCP/IP suite is comprised of two protocols: the TCP and the *User Datagram Protocol* (UDP).

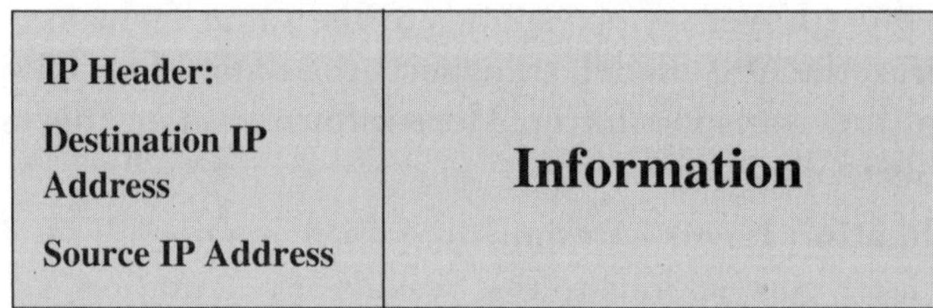

FIGURE 5-3 Datagram structure

The TCP performs the Transport Layer functions of the IP. The TCP is designed to provide for data connection services to support applications. It contains parameters to ensure reliable and error-free delivery of datagrams. The TCP also ensures that the datagrams are delivered in sequence and without missing packets. Assume that an application sends a file to the TCP. The TCP adds a header to the datagram. The datagram is now called a segment. The TCP will receive incoming data from the IP Layer and then determine which application should receive the segment.

The UDP is a connectionless function that is normally used by database lookup applications. The UDP supports stand-alone messages like a simple query. Figure 5-4 depicts the Transport Layer.

IP Protocol Architecture

The TCP/IP protocol suite is comprised of a number of protocols that support a variety of functions such as applications and network management. The following is a list of dominant and popular members of the TCP/IP protocol suite.

- **File Transfer Protocol** FTP supports the transfer of files between computers that are remote from each other.
- **Simple Message Transfer Protocol** SMTP supports electronic mail transmission and reception.
- **Simple Network Management Protocol** SNMP supports network management.

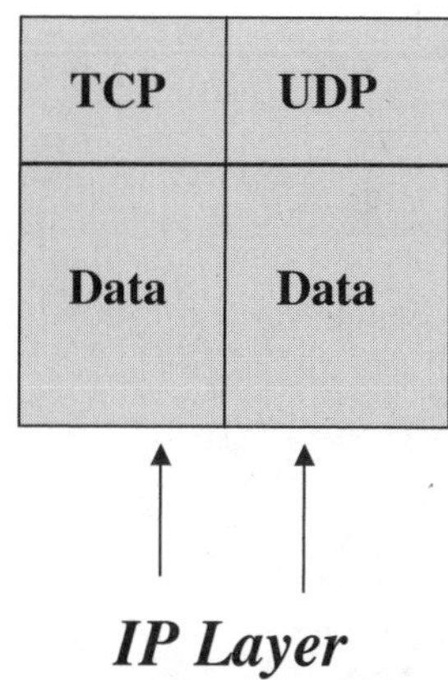
TCP
UDP
Data
Data
IP Layer
The IP Layer will transmit data to the appropriate Transport Layer sub-layer.

FIGURE 5-4 Transport Layer TCP

- **Transmission Control Protocol** (TCP).
- **User Datagram Protocol** UDP is normally bundled with the IP. UDP supports connectionless transmission.
- **Internet Control Message Protocol** ICMP supports diagnostic functions.
- **Internet Group Management Protocol** IGMP supports group management on a router.
- **Routing Information Protocol** RIP is a popular routing protocol in use today.

These protocols are all part of the TCP/IP protocol suite. The following diagram highlights the layering of these protocols in a way that enables them to complement one another. See Figure 5-5.

The TCP/IP protocol suite was designed to interconnect dissimilar network elements and networks. In other words, one could be looking at a series of LANs or *Wide Area Networks* (WANs) interconnected together. To the wireline or wireless telecommunications engineer, it would look a like a jumble of networks hobbled together. TCP/IP enabled the interconnection of private networks, which ultimately led to the Internet. Early in the life of the TCP/IP protocol suite, these private net-

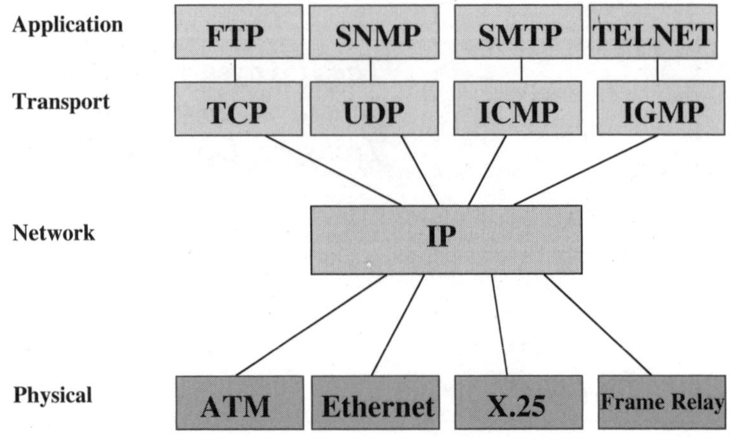

FIGURE 5-5 The TCP/IP protocol: Layered with multiple protocols

works had been stand-alone networks that performed a variety of different tasks, ranging from academic to business functions. The beauty of TCP/IP was its ability to enable this jumble of networks to communicate with one another. The physical translation of the protocol architecture is the series of hosts and routers used to bring the Internet to life. The router, like the tandem in the voice telecommunications world, is the central piece of switching equipment that enables the interconnection of these dissimilar networks.

PHILOSOPHICAL DIFFERENCE BETWEEN THE INTERNET AND NON-INTERNET WORLD

The Internet stresses flexibility, not reliability. In a traditional telecommunications environment, which encompasses the wireline and wireless carriers, the stress is on reliability. Due to this stress on reliability, the wireline and wireless carriers maintain an exhaustive set of operating and technical requirements, which govern what can be interconnected into their networks. During the early 1990s, a large wireline carrier suffered a massive network failure affecting multiple states, resulting in a loss of service for several hours. The outage could not have been anticipated even with the exhaustive simulation tools available at the time. ISPs have suffered network outages that have lasted hours. However, in the case of the wireline network, one hour of outage is unacceptable. This incident was resolved quickly by the wireline carrier, but not without public and industry repercussions. In the traditional telecommunications environment, if a piece of equipment does not meet specific operating and technical standards, then it will be prohibited from use in the network. The Internet, however, stresses flexibility; therefore, the need for stringent standards does not exist. The Internet players' concentration on flexibility has enabled multiple equipment vendors to manufacture a variety of computers, hosts, routers, storage devices, laptop computers, software utilities, and network management tools without having to account for national standards of performance. If an Internet company's product does not satisfy the marketplace, it usually goes out of business.

By looking at how the Internet is configured with multiple network types and multiple configurations, it is difficult not to be amazed at the Internet's ability to adapt. Figure 5-6 is an illustration of the role of the router and the hosts in the *Inter*connection of the *net*works.

The IP is a "best effort" protocol. The focus on the moving of data in the Internet world is on the flexibility of interconnection, not the reliability of the data at the destination point. In other words, the only guarantee of the Internet is that a "best effort" will be made to ensure the data arrives intact. Many "Internet bred" telecommunications pundits bask in the Internet's ability to enable the communication of so many different devices quickly. However, these same industry pundits discount the voice telecommunications experts need to ensure data integrity (bit error rates on the order of one in every billion) as overkill. In the world of banking, making claims of anything less than one bit error in every billion bits is suicidal. Many of the current Internet players are early pioneers, so the view of the world is focused on technology. However, one should remember the focus is on the customer, not the technology.

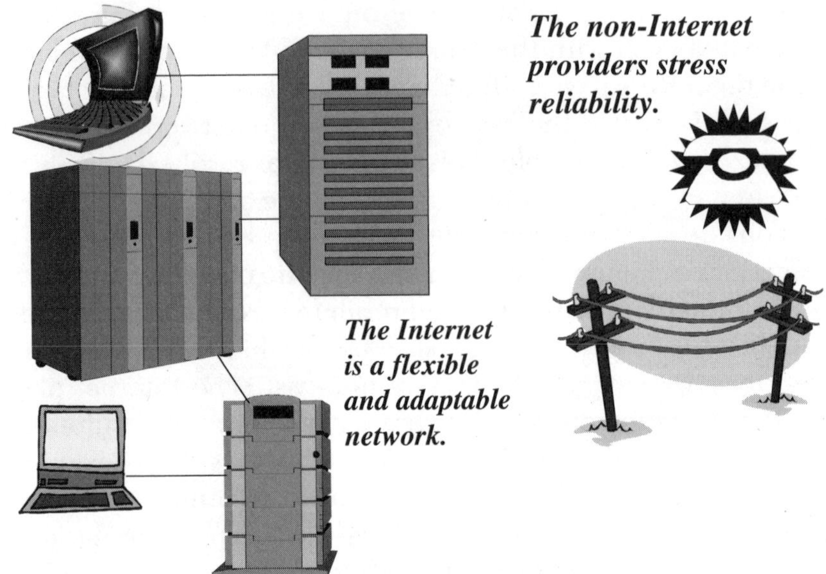

FIGURE 5-6 The Internet – Flexibility

INTERNET PROTOCOL — VERSION 4 VERSUS VERSION 6

The original intent of the Internet (inter-network) as defined by the Department of Defense had been to support the telecommunications needs of the government and the military. The original concept of the Internet made sense especially during the Cold War. From a national security perspective, the Internet continues to be applicable. One cannot depend solely on the wireline and wireless carriers for telecommunications needs. Alternative arrangements are prudent steps to take. Conceptually, the Internet has not changed its requirements or capabilities. Figure 5-7 illustrates the capabilities of the Internet. The Internet is capable of supporting the following:

- A multitude of computers, hosts, routers, and other devices made by multiple vendors.
- Network growth without regard to end-to-end planning.

- Network growth without affecting the (public) Internet.
- The implementation and deployment of all types of subscriber-based applications without affecting the various networks interconnected to the Internet.
- Operational flexibility. Maintain operation regardless of the network health/status of any interconnected networks.
- Backward and forward compatibility. This means more than just network growth. The ability to maintain communication between network types, regardless of the software or hardware versions in place, is an enormous benefit to Internet users.

INTERNET PROTOCOL (IP) VERSION 4

The IP currently in use today (year 2000) is called *Internet Protocol Version 4* (IPv4). To the layman, IPv4 is simply called IP. However, IPv4 has encountered a number of challenges. To those people who only know the Internet world, these chal-

The Internet:

- ***Is Adaptable.***
- ***Can support a multitude of terminal devices.***
- ***Can support growth without end-to-end network planning.***
- ***Can support multiple user applications.***
- ***Can support back and forward compatibility.***
- ***Is flexible.***

FIGURE 5-7 Internet capabilities

lenges or problems are viewed as only perceived challenges or problems. To those in the wireline and wireless telecommunications worlds, these challenges or problems are not perceived, but real. As the multitude of carriers work to bring voice and other services to the residential and business user, the following issues have come to light. IPv4 is viewed as limited in the following areas:

- **Security** Encryption and authentication are not required in IPv4 implementations. Internet users and MIS departments will implement their own security measures, which may result in the erection of simple firewalls. Some basic security measures can be executed. More on this subject will appear later.
- **Reliability** Reliability is a subjective measure of dependability. Reliability is one of those soft, "touchy-feely" words that means different things to different people. As defined in Chapter 4, "Internet and Technology Service: Provisioning in the Information Marketplace,": "Reliability is very different than either availability or MTBF. Reliability is a measure of performance or dependability, whereas the system is operational. It is a subjective measure that takes into account availability, MTBF, and consistent quality of product. Reliability is a subjective measurement because how much weight is placed over one measurement compared to another is dependent on the person." IPv4 does not maintain a set of control protocols for network management purposes.
- **Quality of Service (QoS)** Service quality is another subjective term that can be measured using objective tools. An Internet carrier should establish specific levels of service quality, whether it is fair, good or bad, or levels one through 10. IPv4 is good for some applications although not good for others. IPv4 has difficulty supporting real time data transmissions like voice. It supports buffering of data without regard to the need for real time interactions, which is fine for non-voice data applications. Unfortunately, IPv4

has no way of being stopped from buffering data or taking network paths that add to the time delays. Under the current IPv4 schema data, voice data and non-voice data are treated the same.

- **Addressing schemes** Addressing schemes refer to the terminal device identifications. In the wireline and wireless telecommunications world, the addressing scheme is called the telephone number and mobile identification number. IPv4 only supports a 4-byte (32-bit) addressing scheme, which means that up to 4.2 billion terminal devices can be supported. If one believes in the strong future of the Internet, he/she can visualize a worldwide shortage of addresses for Internet terminal devices.

The current IP is not a telecommunications protocol like SS7 or MF signaling. SS7 and MF signaling have strict operating parameters under which they operate. They maintain strict procedures for setting up, transmitting, and tearing down calls or low-speed data calls. If these rules are not adhered to, the communication (call) is either not set up or is torn down in the middle of the communication (call). In comparison, IPv4 is a forgiving protocol. It enables a network manager to maintain an established transmission link with "spit and baling wire." However, IPv4 does not guarantee delivery of any datagram. An Internet network manager can only guarantee that a datagram is launched. As indicated previously, an Internet company has no way of guaranteeing the arrival of all the datagrams. Furthermore, one cannot even guarantee that the delivered information can even be assembled in the correct order. Figure 5-8 represents IPv4's limitations.

None of what has been stated should be construed as a negative review of IP. In fact, it is IP's flexibility and adaptability that has facilitated the growth of the Internet. Specifically, four features exist: fragmentation, reassembly, encryption, and adaptive/dynamic routing.

Limitations

Security

Reliability

Quality of Service

Addressing Schemes

FIGURE 5-8 IPv4's limitations

FRAGMENTATION AND REASSEMBLY Datagrams can vary in size. As a result of this size variation, it is possible that the datagram can get stuck in an intermediate network while in transit to its destination. Routers are preset to accept datagrams of a specific size. However, in order to ensure that the information is not lost, the router in the intermediate network will chop the datagram into smaller pieces.

The destination router takes the responsibility of reassembling all of the pieces. This capability ensures that no matter what operational parameter may have been established by another network manager, the datagram still continues its journey. It is unfortunate that the fragmentation and reassembly capability drastically increases the amount of time it takes for the ultimate destination to obtain the full datagram.

ADAPTIVE/DYNAMIC ROUTING Adaptive/dynamic routing refers to the IP ability to determine the best path of transmission in real time. In other words, as the datagrams are traversing the network in a multitude of paths, every network router is making a decision (in real time) as to the next best intermediate destination of the datagram. This means that any physical change in one of the many intermediate networks would not affect the datagram. In other words, a change in any one of the networks would cause the datagram to take a different route. The capability of real time adapt routes is a fantastic capability that serves to increase the flexibility and usefulness of IP. Figure 5-9 is an illustration of the IPv4 datagram.

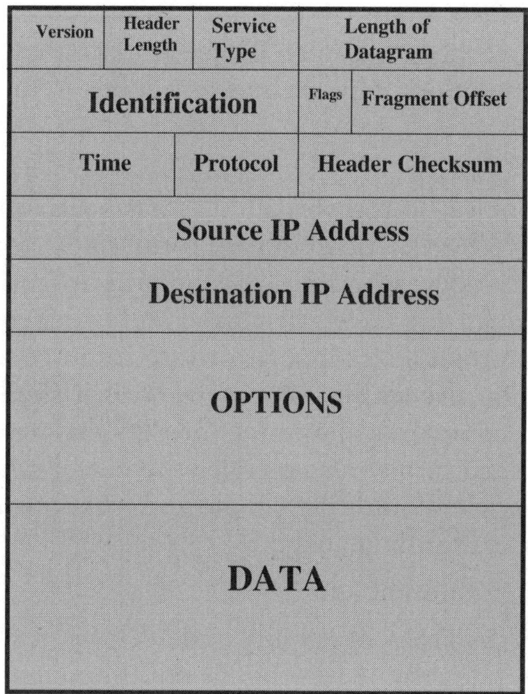

FIGURE 5-9 IPv4 datagram

IPv4 Summary

A number of weaknesses and issues exist concerning IPv4's ability to support time sensitive applications like voice. However, these limitations do not diminish the power of the protocol in any way. If IPv4 was not as flexible in its ability to support the interconnection of multiple vendors devices and in its ability to adaptively route, the Internet would not exist. The next section will take a brief look at *Internet Protocol Version 6* (IPv6).

Internet Protocol (IP) Version 6

Many industry pundits are claiming to find salvation with the arrival of IPv6. However, one should note that in the world of telecommunications (of which I claim the Internet is a part), nothing is static. IPv6 is still under development by the *Internet*

Engineering Task Force (IETF) and various industry consortiums. So much hope has been placed in IPv6 that many call it "IP Next Generation." IPv6 brings a number of innovations to the Internet:

- The current 32-bit (8-byte) address has been expanded to 128 bits (16 byte). The expanded format now enables hierarchical addressing, which is helpful in routing schemas requiring more structured and faster routing. The IPv6 format also supports private Internet site addresses for use within a company. This is similar to a telephone company's allowance for PBX dialing extensions. The expanded format also enables the use of multicast addressing where multiple people and locations receive information simultaneously.
- Self-configuration of addresses.
- Enhances the protocol's ability to detect the next-hop routers and determine whether or not the routers are active, which can reduce latencies arising from transit times.
- Enhances the protocol's ability to detect a healthy and usable transmission path.
- The addresses can now be structured to simplify address delegation and routing. The addressing methods used in IPv6 are similar to bulk addressing of signaling points in a SS7 network, where the nodes can be addressed as either a cluster of nodes or as individual nodes. This addressing methodology simplifies routing plans.
- Supports a new flow label that effectively introduces the concept of information flows, which can be used to support voice and maybe even video. Packets can now be tagged with just the origin and destination so the packet does not need to have its entire header read, just the tag.
- Authentication.
- Data integrity.

- Encryption can now be supported.
- Extension headers are introduced to simplify the main header.
- Improved congestion control.
- Enhances the ability to encapsulate and transport other protocols.

The changes in the IP have gone as far as changing a great deal of the terminology. The most significant change is the change from "datagram" to "packet." Currently, datagrams only transport information payloads to support the IP. However, in IPv6, the datagrams can carry payloads from other protocols; hence, the more general term of "packet" is used. Figure 5-10 is an illustration of the IPv6 packet.

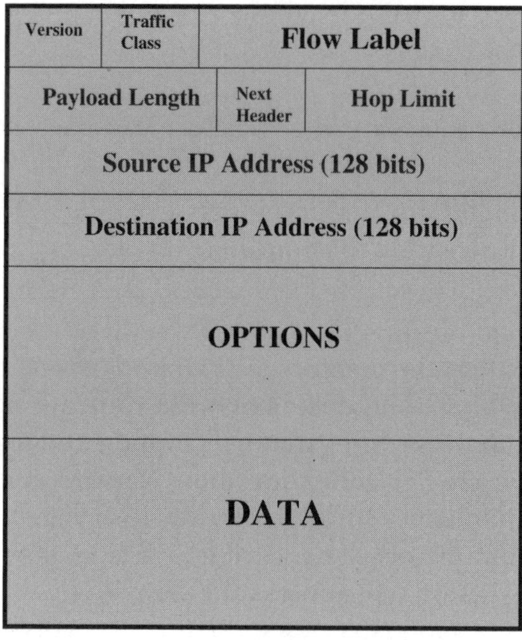

FIGURE 5-10 IPv6 packet structure

TRANSITIONING FROM IPV4 TO IPV6

This book does not intend to examine the details of IPv6. IPv6 is a work in progress. The important detail to understand is that the IP is undergoing changes that enable it to support data and voice services. However, the arrival of IPv6 does not mean that the entire Internet community will change overnight. In fact, the transition to IPv6 will be filled with challenges and troubles. The new addressing format will be the source of almost all of the issues.

Networks of all kinds face the same issues when a protocol has been updated. Current client (user applications) applications perform transactions using 32-bit (4-byte) addresses. Clients, like a telephone or wireless switch, expect to see an address (or telephone number) in a specific format. When a different format is used without properly making the appropriate software changes, the clients will not be able to perform their functions. Therefore, it becomes imperative to upgrade all of the applications in use. The changes will be gradual for the following reasons:

- The cost of converting in a "flashcut" manner would be prohibitive.
- Millions of clients exist in the network today; therefore, it would be impossible to coordinate any kind of flashcut.

For a period of time, IPv4 and IPv6 will co-exist in the network. Fortunately, the creators of IPv6 understood the practical realities of industry and customer-wide software upgrades and created a set of procedures that will support the use of address converters. Figure 5-11 illustrates how Network A and Network B will communicate with one another if one network is IPv6 operational and the other network is still IPv4 operational.

The next section will address the new wireless protocols. At this moment, two industry news-making protocols are available. This book will describe the protocols; however, for the purposes

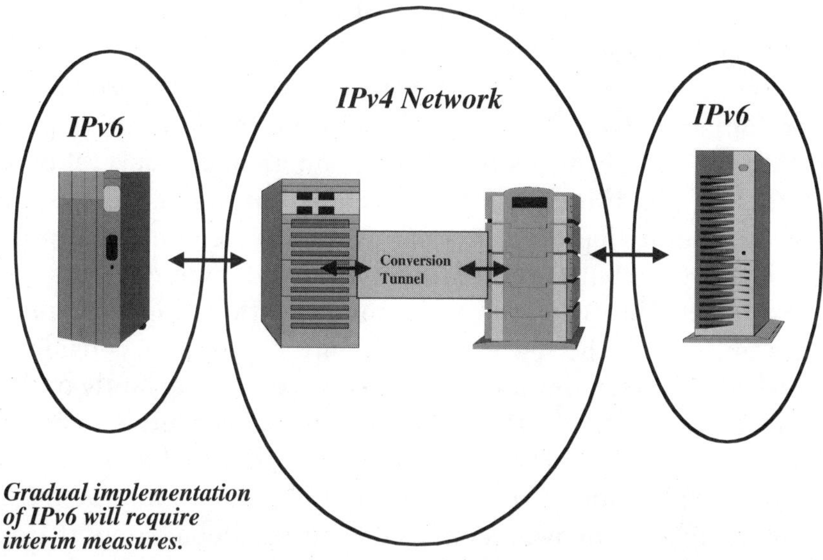
IPv6
IPv4 Network
IPv6
Conversion
Tunnel
Gradual implementation
of IPv6 will require
interim measures.

FIGURE 5-11 Interworking IPv4 and IPv6 networks

of this book, a keener interest in the business implications of the protocols will be taken.

WIRELESS INTERNET PROTOCOLS

The wireless industry is going through a major technology and consumer product shift. Until 1999, the wireless industry in the United States was considered a growing industry that was going from exponential growth to a steady, but less dramatic growth. Being able to make or receive calls at anytime was one of the drivers in wireless growth in the United States. Throughout the rest of the world, the wireless industry is growing for a number of different reasons ranging from convenience to a novelty to primary communication needs. During 1999, the entire global Internet community was all the rage on the financial markets. The Internet was an untapped market for the wireless community. The reader should remember that the Internet is an information medium; therefore, it is natural for the wireless carrier to take advantage of this medium.

Two principal wireless Internet protocols are under development. The *Wireless Application Protocol* (WAP) is being developed in North America, South America, and several other countries, and iMode is being developed in Japan. Both protocols are being deployed in some fashion as commercial products. However, both are still undergoing development work. For those unfamiliar with the telecommunications industry, products tend to hit the marketplace before anything is considered finished. The telecommunications business is constantly undergoing change. The WAP is being developed via a large, industry-wide consortium called the WAP Forum that includes wireless carriers and vendors from several nations. One must be reluctant to say that the protocol has definitive, global support from all carriers and manufacturers. The iMode protocol is a company proprietary protocol being developed by NTT DoCoMo, yet NTT DoCoMo is a member of the WAP Forum. Competition and the need for profits have a way of changing the landscape of cooperation.

This chapter will not espouse a position of which protocol is better or worse. However, it will look at the differences from a customer perspective. Customers do not care how the product works. They just care that the product works well, is usable, and is "cool to use." Services will be addressed in Chapter 6, "E-Commerce and Mobile E-Commerce."

WIRELESS APPLICATION PROTOCOL (WAP)

The WAP is the result of an ongoing effort to define an industry-wide standard for developing data and Internet-based applications that operate over wireless telecommunications networks. This effort takes the form of an industry forum known as the WAP Forum.

The WAP Forum is comprised of nearly 500 members, which include switching manufacturers, computer manufacturers, terminal device manufacturers, content providers, service providers, and software developers. The membership list is impressive. To enable operators and manufacturers to meet the needs of highly dynamic markets, the WAP (protocol) was designed in a layered fashion. In other words, the protocol model is similar to the OSI model in that each functional layer supports a discrete set of functions. As previously mentioned, the OSI model consists of seven layers. The communication functions are broken down into a hierarchical set of layers. Each layer performs a related subset of the functions required to communicate with another system. Each layer also relies on the next lower layer to perform primitive functions and conceal the details of those functions. It also provides services to the next higher layer. The layers are defined in such a manner that changes in one layer do not require changes in other layers. By partitioning the communication functions into layers, the complexity of the protocol becomes manageable. In the case of the wireless Internet, the complexity of delivering services is enormous due to the mobile feature of the customer. Figure 5-12 illustrates the WAP reference model. The protocol is somewhat similar to the OSI model. Figure 5-13 is a comparison of the two models.

WAP is completely new to the telecommunications industry. Unfortunately WAP is not compatible with the current wireline IP (IPv4 or IPv6) or any of the wireline Internet Web site scripting languages. This is not entirely bad or good. WAP has proven that wireless is so popular that interoperability is not needed at this time. However, interoperability will be needed if the full power of the Internet is to be realized. More information on interoperability will be discussed later. WAP has two goals:

- To enable wireless users to easily access and interact with information and related services.
- Access the Internet using a wireless handset.

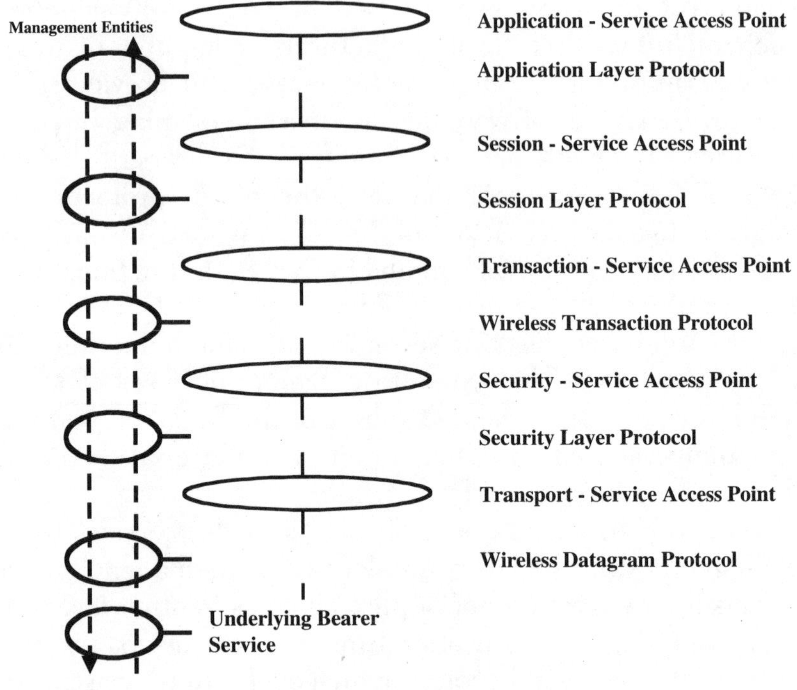

FIGURE 5-12 WAP reference model

The WAP Forum was created with the following objectives in mind:

- To create a protocol specification to bring Internet content and data services to the digital wireless telecommunications users.
- To create a protocol specification that works across multiple wireless technologies.
- To enable the creation of applications and content that scales across a wide range of networks and devices.

As this chapter is being written, a business war is being waged across the oceans, domestically, across technologies, and across industry segments for domination of the Internet busi-

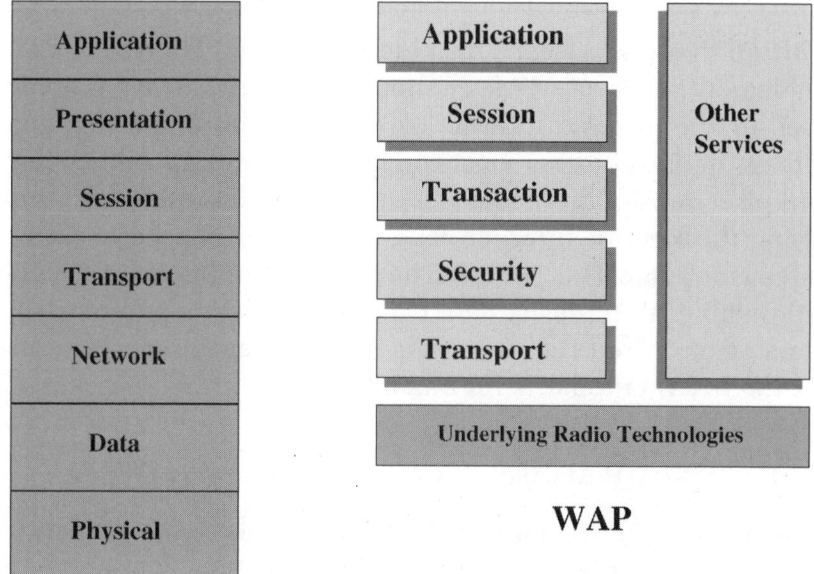

FIGURE 5-13 WAP model versus OSI model

ness space. This war is being waged on both a wireless and wireline front. The wireless carrier community faces the challenge that the Internet was designed for the wireline world.

The wireline world's version of the Internet is comprised of the following:

- Desktop computers
- Laptop computers
- High-bandwidth transmission
- Larger displays on the computers
- Heavy use of graphics
- Interactive icons
- Interactive graphics
- Streaming video
- Streaming audio

The wireless world began its protocol development effort with an eye toward early commercialization. Therefore, early implementations of WAP cannot support the heavy graphics used by the wireline Internet. The real issues of small display screens, mono color screens, and limited video capabilities (limited pixel density) in the handset prevented handsets from supplying the kind of Internet activity the user is used to on the wired computer. This situation does not mean Internet access is unavailable. WAP devices are capable of sending and receiving e-mail from wired Internet users. Remember the principal use for the Internet today is for e-mail.

WAP and Non-WAP Architectural Comparisons (WWW)

The openness to the individual software programmer (amateur and professional) for the purposes of Web site creation or content access distinguishes the Internet. The *World Wide Web* (WWW) is the killer user application of the Internet. It is the physical reality of multiple hosts connected and communicating with one another. The WWW is an Internet application that enables the user to exchange documents, access servers owned by multiple numbers of people, use multiple viewing formats, access the Web, and manipulate the content using any one of a multitude of applications.

The *World Wide Web Consortium* (W3C) is a global industry-wide voluntary organization of companies and individuals dedicated to creating and maintaining an environment that is supportive of the growth of the Internet. Specifically, the W3C is promoting the following programming model represented in Figure 5-14.

The W3C has created a set of standards that has enabled software developers to write applications and create content so that they are presented in a set of standard formats. These standards have also enabled the creation of the Web browser. The Web browser is an application that enables a user to send out requests for information (data queries) to any Web site. The Web browser is capable of supporting the access of data from

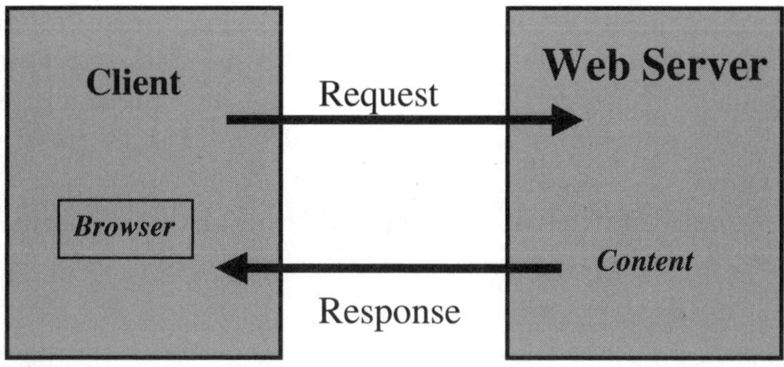

FIGURE 5-14 WWW programming model

multiple information sources. The W3C has promoted a series of Application Layer tools that ride over the Internet protocol. Many of these popular tools are focused on the presentation of the data. These are some of the tools:

Hypertext Markup Language **(HTML)** HTML is a language that describes how a document is to be displayed/presented to the user.

Hypertext Transfer Protocol **(HTTP)** HTTP is a protocol that defines how the Web client and a Web server communicate with one another.

Standard Generalized Markup Language **(SGML)**

SGML is a language that describes elements in portable documents.

Extensible Markup Language **(XML)** XML is a subset of SGML. XML facilitates how new document formats are defined; that is, XML describes the data itself. XML is a vendor neutral format.

Document Object Model **(DOM)** DOM provides a standard object model for the presentation of HTML and XML documents. The DOM enables programs and document scripts to access and update the structure, content, and style of documents.

If it were not for the work of the W3C, the Internet would simply be a world of e-mail and visually unexciting, interactive Web pages. The way in which the W3C has approached the WWW concept is open for growth. Figure 5-15 is an illustration of the Web.

In general, the W3C has created a series of standards that support the following:

Naming Models All servers and Web content are named with a *Uniform Resource Location* (URL), *Uniform Resource Name* (URN), or *Uniform Resource Identifier* (URI). These naming models are unifying concepts that provide servers and content names, tell where the various items are located, and tell how to get the items (identifies the protocol needed to get the items). The URL is the underlying naming model. An example of the URL would be http//servername/filename/html. As the reader can see, the URL starts by using the access protocol, followed by a

FIGURE 5-15 The WWW

sentence delimiter, followed by the server, followed by the name, and lastly followed by the resource to be accessed to view the item.

Content Typing A certain type identifies all content on the Web. This content type identification enables Web browsers to process content appropriately.

Content Formats All content must be presented in a format that can be understood by the various Web sites. HTML and JavaScript are such formats. JavaScript is the market leader in scripting languages and a benchmark for scripting languages.

Standard Protocols The Internet protocol suite supports a variety of application protocols. The application protocol called HTTP is the most commonly used transaction and session protocol.

The WWW is more than just an Internet application. The WWW has become a software template for the concept of distributed computing. The WWW has become a software infrastructure programming model for the industry. This use of the term "infrastructure" is in recognition of the foundational role of the WWW work. The infrastructure model is depicted in Figure 5-16.

The model has defined three classes of servers. These classes are as follows (see Figure 5-17):

Origin Server The Origin server is the home of the user.

Proxy Server The Proxy server is an intermediary program that acts as a client and server in those instances where direct communication between the server and client are not possible.

Gateway Server The Gateway server is like a Proxy server in that it makes requests. However, a Gateway server acts as if it is the Origin server. The Gateway server is a face for Origin servers.

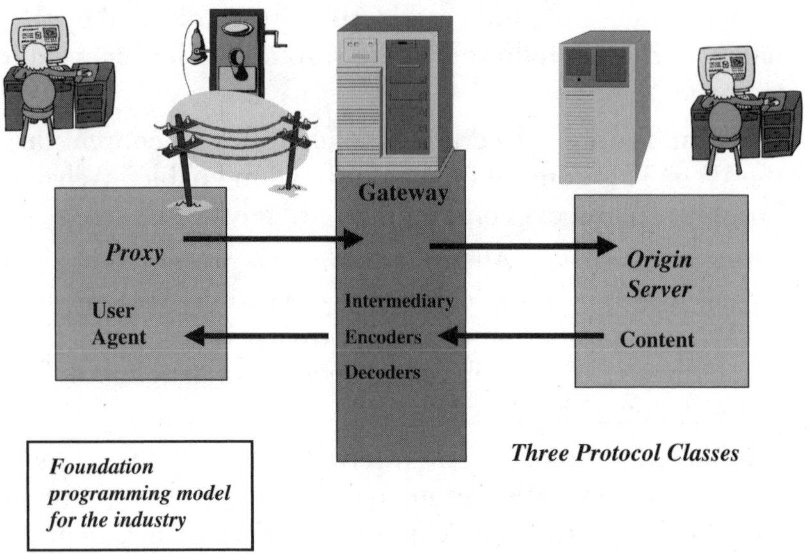

FIGURE 5-16 Infrastructure model

Infrastructure Model

Origin server – The user

Proxy server – Intermediary program, where direct communication between server and client are not possible.

Gateway server – Intermediary program that acts as both Origin server and Proxy server.

The Infrastructure model serves as a basis for programming.

The work of the W3C has served as a lighthouse for the Internet industry.

FIGURE 5-17 Server classes

The WWW work has served as a basis for all wireless Internet work to date. Like the WWW, the wireless community has created a series of models and tools through which the wireless communities' needs are met. The WAP software infrastructure programming model looks similar to the one used in the WWW work. See Figure 5-18.

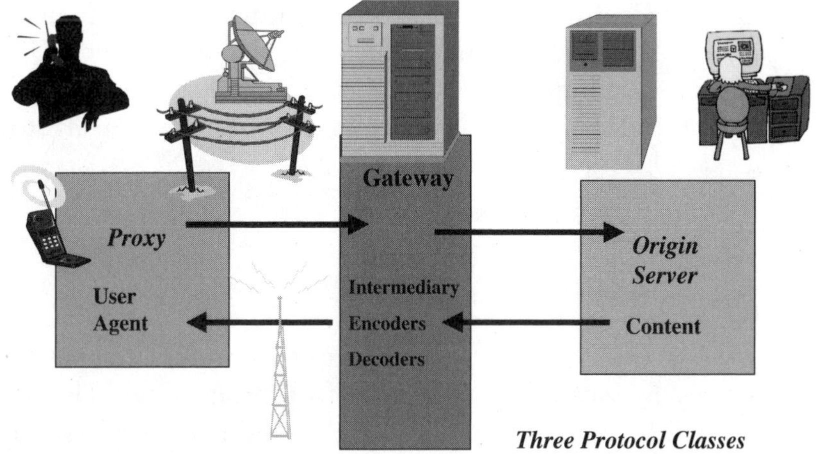

FIGURE 5-18 WAP software infrastructure programming model

The WAP Forum has defined a similar set of programming mechanisms. These mechanisms or conventions are based on the WWW work. These mechanisms are as follows:

Naming Models All servers and Web content are named with a WWW URL.

Content Typing A certain type identifies all content on the Web. This content type identification enables Web browsers to process content appropriately.

Content Formats All content must be presented in a format that can be understood by the various Web sites. HTML is a format used in the WWW. XML is also used in the WWW; however, as of 2000, XML has not reached a level of predominance. Due to the unique physical attributes of wireless handsets, the WAP supports a format called *Wireless Markup Language* (WML). WML is designed to support the compressed data needs of the wireless network. This use of the word "compressed" addresses the small information pipe nature of the wireless handset.

Standard Protocols The IP suite supports a variety of application protocols. The application protocol called HTTP is the most commonly used transaction and session protocol. In the case of the wireless community, HTTP 1.1 is used, which is based on HTTP.

WAP AND NON-WAP TERMINOLOGY

The various Internet programming and document format languages deserve further attention before proceeding. This discussion will focus on the main topics of industry conversation.

Hypertext Markup Language (HTML) is the universal language of the WWW. It was first introduced in 1995. The current version, in the year 2000, is HTML version 4.0. HTML tends to function or behave differently with different browsers. The occurrence of differences is normal for any standard that is being interpreted by a wide variety of vendors.

Extensible Markup Language (XML) is a new technology for the WWW. The W3C created XML to simplify business transaction on the Web. HTML is used today; however, it supports communication between the user and the computer. HTML tags are used to support browsing. HTML tags describe the presentation of the data. See Figure 5-19 for a rendering of an HTML tag.

In the case of XML, the tag describes the kind of information that is presented in the data. XML enables the programmer and the computer to understand what the data contains. Furthermore, whereas HTML tags are predefined and therefore unchangeable, the XML tags are definable by a developer. Therefore, any data item can be identified. This also means the Web pages can function like database records. The data item defined could include the followiing:

- Product type
- Product name
- People's names

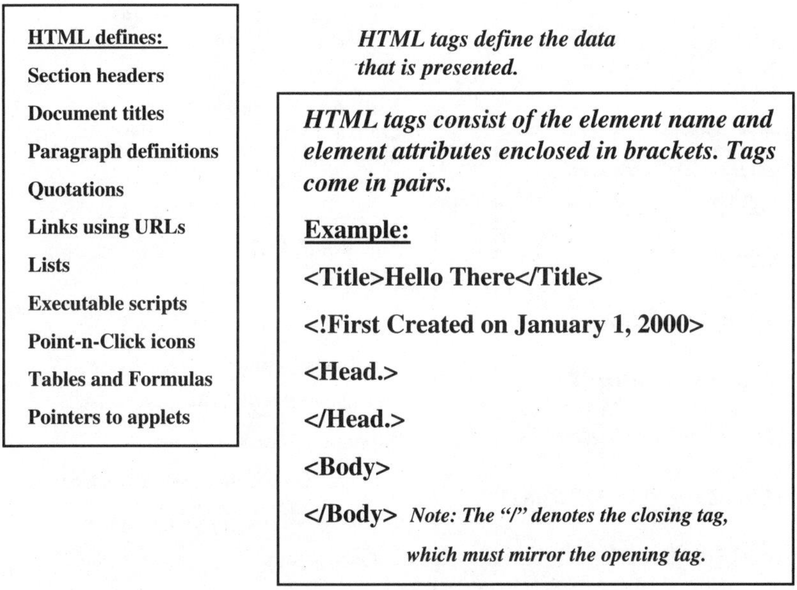

HTML defines:
Section headers
Document titles
Paragraph definitions
Quotations
Links using URLs
Lists
Executable scripts
Point-n-Click icons
Tables and Formulas
Pointers to applets

HTML tags define the data that is presented.

HTML tags consist of the element name and element attributes enclosed in brackets. Tags come in pairs.

Example:

<Title>Hello There</Title>

<!First Created on January 1, 2000>

<Head.>

</Head.>

<Body>

</Body> ***Note: The "/" denotes the closing tag, which must mirror the opening tag.***

FIGURE 5-19 HTML tag

- Price of product
- Final product sales price
- Amount of money wed
- Directories of files

XML is a rigidly defined language that has caused the establishment of a common method of identifying data. XML has been able to support business-to-business transactions. XML will likely become the defacto standard for data interchange. Figure 5-20 is a rendering of a XML tag.

Given the fact that most companies and even organizations within the same company do not use the same set of software tools, XML can serve as a way of bridging the communication gap. XML supports business-to-business communications and will be able to support the creation of "smart agents." A smart agent (intelligent software application) interprets information and then acts on the information based on the smart agent's

XML tags are definable by the programmer.

HTML tags are geared towards supporting Web browsing by the human user.

XML tags enable programmers to granularize and customize the data definitions.

XML tags define items such as:

Product Type

Product Name

People's names

Price of Product

Directories of Files

Money owed

XML Sample Code:

```
<address>
<name>
<title>Doctor</title>
<first-name>PJ</first-name>
<last-name>Doe</last-name>
<street-name>1 Main Street</street-name>
<city-name>Mayberry</city-name>
<state>NC</state>
<zipcode>11111</zipcode>
```

FIGURE 5-20 XML tag

understanding of the data. XML can make it far easier than it is today for the smart agent to understand what the data is. Search engines can also benefit from the use of XML.

XHTML stands for Extensible Hypertext Markup Language. As noted previously, HTML is the universal language of the WWW. It is used to support communication between the user and the computer. XHTML is a combination of HTML and XML. The combined format will define the data and the presentation of the data. Essentially, a programmer would be able to create HTML documents with XML functions combined. This could result in a HTML document with video and music incorporated. Figure 5-21 is a rendering of a XHTML tag.

Many have compared WML to HTML, but this comparison is incorrect. WML is more similar to XML than it is to HTML. The creators of WML, the WAP Forum, will explain that WML is based on XML. WML is designed to support Web transactions on low-bandwidth, small handheld devices (that is, mobile handsets) that have small viewscreens. WML defines the type

XHTML tags are definable by the programmer.

XHTML combines the structure of HTML and flexibility of XML.

XHTML tags enable programmers to granularize and customize the data definitions.

XHTML tags define items such as:

Product Type

Product Name

People's names

Price of Product

Directories of Files

Money owed

XHTML Sample Code:

```
<address>
<name>
<title>Doctor</title>
<first-name>PJ</first-name>
<last-name>Doe</last-name>
<street-name>1 Main Street</street-name>
<city-name>Mayberry</city-name>
<state>NC</state>
<zipcode>11111</zipcode>
<file-type>audio</file-type>
```

FIGURE 5-21 XHTML tag

of data traversing the wireless Internet. Unlike XML, WML must support data transactions that are not only being displayed on small handsets, but also live with limited computing memory in the handset. Figure 5-22 is a rendering of a WML tag.

WHAT ARE WAP'S LIMITATIONS?

At this time in the year 2000, the WAP networks in place are completely independent of the WWW. Although an individual can browse Web sites on his/her wireless handset, he/she can only browse Web sites that are supported by the wireless carrier or its gateway provider. This is not good in the long run for the wireless user.

The WWW is the dominant application of the Internet and should remain the model upon which all Internet work is performed. However, the WAP Forum is working toward converg-

WML is based on XML.

WML does not define processing instructions beyond what is defined by XML.

WML must support transactions that are displayed on small handheld device screens.

WML Sample Code:

```
<address>
<name>
<title>Doctor</title>
<first-name>PJ</first-name>
<last-name>Doe</last-name>
<street-name>1 Main Street</street-name>
<city-name>Mayberry</city-name>
<state>NC</state>
<zipcode>11111</zipcode>
```

FIGURE 5-22 WML tag

ing its work with that of W3C. For now, the WAP in its current form is more than sufficient. In fact, WAP has actually reinvigorated the entire wireless industry. Critics of WAP should remember that little steps are important in the beginning. The WAP in its current form is a significant little step.

WAP PROTOCOL ARCHITECTURE

The WAP Forum has created an architectural schema that supports scalability and flexibility for applications development in the wireless arena. The protocol architecture is a functional layered model. The layered approach enables development work to occur in various areas of the WAP protocol. Note that this is the same approach supported by the OSI model. The WAP architectural model is depicted in Figure 5-23.

The components of the WAP protocol architecture are described below as follows:

WIRELESS APPLICATION ENVIRONMENT (WAE) LAYER The WAE Layer supports a general-purpose development environment for applications. The WAE Layer enables a developer to combine aspects of the WWW technologies and the current

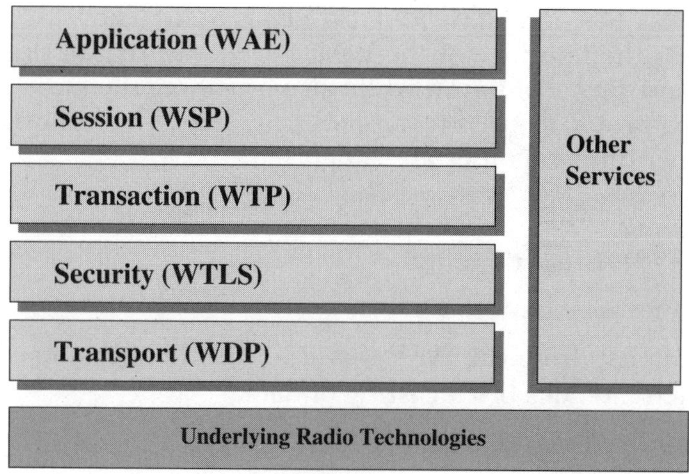

FIGURE 5-23 WAP protocol architecture

wireless voice/data technologies. It is primarily focused on interoperability between the multitude of wireless technologies in use. The WAE Layer contains the Web browser supported by the wireless carriers. The browser in the wireless world is actually called a microbrowser, which emphasizes the fact that the wireless handset is supporting a limited form of Web browsing on the small display screen of the wireless handset. This particular layer supports the following:

- WML
- WMLScript—Similar in function to JavaScript, which is used in the WWW. However, WMLScript is not JavaScript.
- Telephony applications.
- Content formats—Supporting data, images, and text.

It must be noted that WML and WMLScript are not compatible with HTML or JavaScript. WML and WMLScript, respectively, embody the same concepts as HTML and JavaScript, but that is all. A movement exists to converge the efforts of the WAP Forum with those of the W3C.

WIRELESS SESSION PROTOCOL (WSP) LAYER The WSP Layer supports the interface of the WAE Layer with that of the Transport Protocol Layer, while also providing the basic networking protocols used to support browsing. The following browsing functionalities are supported:

- HTTP/1.1 functionality.
- Feature negotiation.
- Data push from the WAP server to the client (handset), which includes broadcasting of data.
- Long-lived sessions.
- Session suspend and resume functionality.

Beyond browsing, the WSP Layer is capable of supporting the WAE in the following ways:

- Establishes the session between the WAP server and the user.
- Tears down the session between the WAP server and the user.
- Negotiates between the user and the WAP server.
- Supports compact encoding of information.
- Supports multiple and simultaneous asynchronous transactions.
- Supports low-bandwidth.

WIRELESS TRANSACTION PROTOCOL (WTP) LAYER The WTP Layer supports the transport mechanism suitable for "thin" clients (wireless handsets) and the low-bandwidth in the wireless network. The use of the term "low-bandwidth" is appropriate when comparing the wireless Internet to that of the wired Internet. The word "transaction" is appropriate because the WTP supports the exchange of information between entities or applications. One example of a transaction would be interactive browsing.

In general, WTP supports the following:

- Improved the reliability of data transfers by ensuring that the number of handshakes and acknowledgments are reduced, which is the concatenation of WTP protocol control information, also known as a PDU. It also supports retransmission of lost data packets. Selective retransmission and flow control are also supported.
- Improves the efficiency of connection-oriented services by eliminating any explicit connection setup or tear down instructions.
- Message-oriented services, excluding streaming services.
- Three classes of transaction services.
  - **Class 0** Unreliable one-way requests. This class of service is intended to support a datagram that is transmitted within the same context of an existing session. This is not intended to serve as a primary method of transmitting datagrams.
  - **Class 1** Reliable one-way requests. This class of service supports a "reliable push" service.
  - **Class 2** Reliable two-way request-reply transactions. This service provides for the normal message transaction where an invocation or request results in some kind of response or reply.
- Optional user-to-use reliability.
- Asynchronous transactions.
- Aborting outstanding transactions.The WTP rides on top of a datagram service.

WIRELESS TRANSPORT LAYER SECURITY (WTLS) LAYER The Wireless Transport Layer Security protocol layer is security protocol based upon the industry standard known as the *Transport Layer Security* (TLS) protocol. WTLS supports WAP over narrow bandwidth channels. WTLS provides the following capabilities:

- Data Integrity.
- Privacy.
- Authentication of the user and terminal device, which supports transactions like credit card transactions.
- Denial-of-service protection, in which the WTLS is capable of detecting and rejecting data that has been replayed or unsuccessfully verified.

Not much else can be said about this layer because it is understood, especially in today's age of hacking and security violations, that security is needed. Security addresses both the privacy and data integrity aspects of the Internet.

WIRELESS DATAGRAM PROTOCOL (WDP) LAYER The *Wireless Datagram Protocol* (WDP) Layer is a Transport Layer protocol. This layer supports communication between the bearer services layer, which is underneath the WDP, and the upper layers of the WAP. The WDP is capable of interfacing with the upper layers in a variety of ways (protocol stacking) in order to execute an application. This is similar to the way in which non-Internet telecommunications services function today. The OSI model, which is used to support networking and the creation of services in the non-Internet world, supports this type of stacking of protocols.

Like the other protocol stacks in WAP, this one is also designed to support narrow bandwidth applications. The WDP is capable of supporting data capable bearer services that in turn are supported by a variety of radio interfaces. In other words, WAP can be used to support radio technologies like *Group Special Mobile/Global System for Mobile Communications* (GSM), TDMA, and CDMA. See Figure 5-24 for an illustration of the WDP's relationship with the bearer service and the radio technology.

The WDP is capable of mapping itself to bearer services in such a way that the bearer services' specific characteristics are

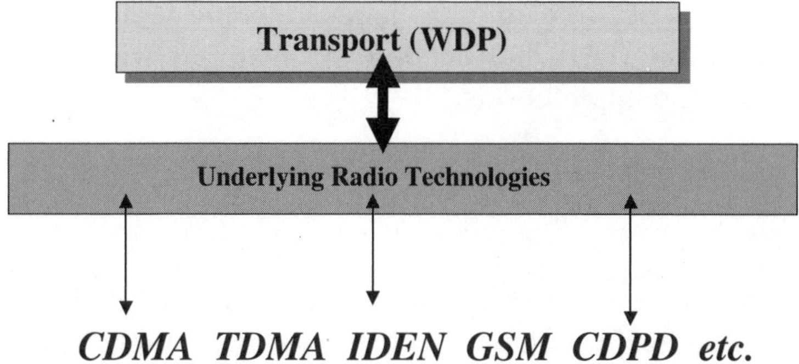

FIGURE 5-24 WDP architecture

optimally addressed. Therefore, the protocol's performance will vary accordingly. This also means that this layer is responsible for communication with physical devices or terminals. Within the WDP Layer is another protocol layer called the Adaptation Layer. The Adaptation Layer is the layer of the WDP that maps the WDP directly onto the bearer service.

The WDP Layer is a highly flexible and adaptable protocol layer that is capable of supporting a bearer service like *Cellular Digital Packet Data* (CDPD). CDPD is an existing service provided by the cellular carriers over separate cellular data networks. The WDP enables the cellular carriers to continue providing existing services over WAP networks. Even more importantly, the WAP is designed to support all of the radio technologies in the marketplace. This flexibility adds to the financial attractiveness of WAP technology.

BEARER SERVICES

The range of bearer services supported by WAP is quite broad. WAP's list of services that are supported include short messaging services, circuit switched data services, and packet data services. WAP is designed to support the varying levels of performance and operating requirements of all bearer services.

WAP has become so entrenched in some technical circles that the term "WAP" has become synonymous with the words "products and applications."

iMode

Many people claim that iMode is similar to WAP, which is true from a basic marketing perspective. Data and Internet access is supported by iMode just as it is by WAP. However, the most glaring difference is that iMode was designed to interface with the WWW. The iMode protocol is a proprietary protocol and service provided by NTT DoCoMo; therefore, little can be written of its protocol structure or the protocol's functional layers. However, the public and market evidence of how the system performs can be addressed.

The most obvious difference to a user is that an iMode handset displays graphics. Now graphics do not "an Internet make." The important concept is that the NTT DoCoMo iMode portal site serves as the gateway to other iMode Web sites or non-imode sites. The iMode service is compatible with the technologies supported by W3C.

The imode handset has a display just as small as a non-iMode handset. However, the iMode portal strips out the unusable data from a WWW Web page and presents it to the user in a coherent format. In order to realize this, the iMode service provider asks the user to provide his/her user profile. The user profile enables the iMode service provider to facilitate access to WWW IP Web sites. One may ask what is the difference between WAP and iMode? The WAP service provider can do the same activity. However, the WAP service provider does not currently support this kind of portal service. The challenge for a WAP service provider is that even if a portal of this kind were provided, the current scripting and markup languages used by WAP are not compatible with the WWW's scripting and markup languages; therefore, stripping out WWW IP Web site pages will be difficult, if not impossible. Figure 5-25 is a rendering of how the iMode network interfaces with the WWW IP network.

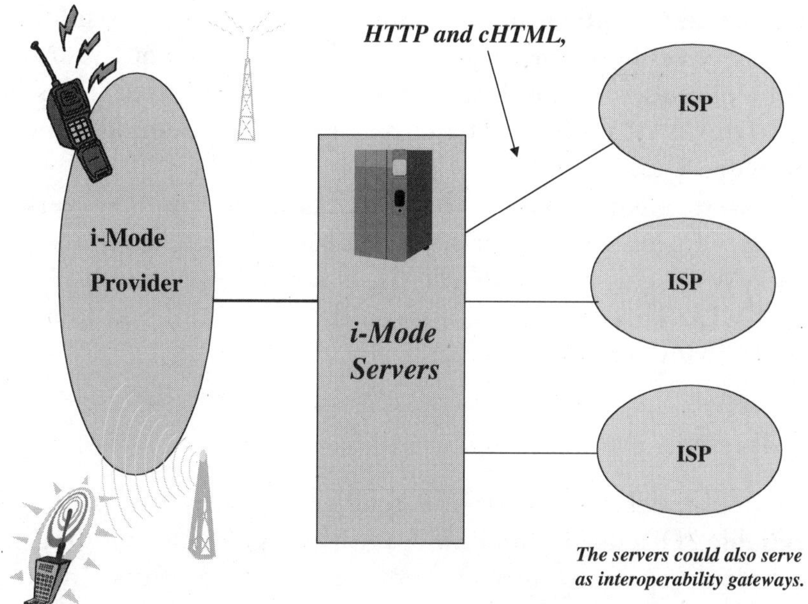

FIGURE 5-25 iMode interconnection with the WWW IP network

The iMode network supports push technology like the WAP network. Both support similar services such as travel, directory assistance, and concierge/directional services. The kinds of services provided have differences, which will be covered in Chapter 6, "E-Commerce and Mobile E-Commerce." However, the iMode network supports interconnection with the larger wired WWW Internet far more easily than the WAP network. This does not mean that the WAP network will not easily support interconnection with the WWW Internet. In reality, the market forces will drive network(s) evolution.

iMode PROTOCOL STRUCTURE

Little can be said of this subject. One can only assume that the protocol model follows the functional layer approach of the OSI model. It makes sense that the NTT DoCoMo approach follows a layered approach. If the iMode model were not layered, it would make applications development nearly impossible and definitely economically unfeasible. The specifics of the iMode protocol model are unclear, but one can assume that it is similar to the WWW IP model and may resemble the model shown in Figure 5-26.

Based on information that has been presented in various industry forms, iMode supports a subset of the HTML used by the WWW. This subset of HTML is known as *compact HTML* (cHTML). This makes sense given the physical limitations of a handset viewscreen.

WAP VERSUS iMODE

The technical merits of one technology over the other are arguable. Discussing technology with engineers regarding the technical aspects of any technology seems pointless. In other words, one should not attempt to convince another of the mer-

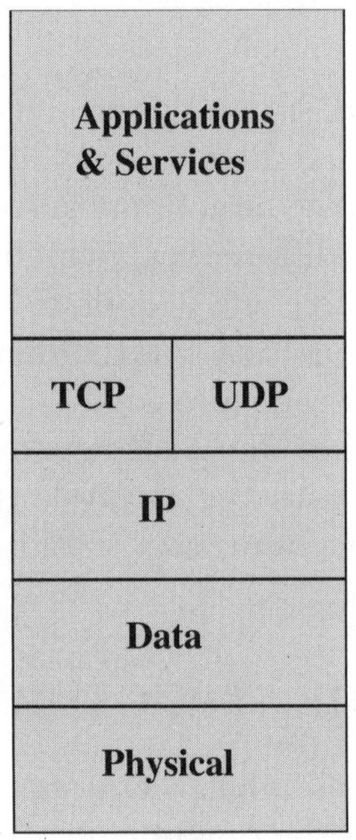

FIGURE 5-26 IP model

its of another religion. However, the merits of the technologies will be discussed from a business perspective, which is a more realistic approach.

One should consider a number of issues when looking at the value of any technology as it relates to its commercial value:

- Does the product have commercial value?
- Is the technology ready for commercialization?
- Does the product do something useful?
- Does the product facilitate an activity that makes a paying customer's life simpler or easier?
- Does the service provider know which market segment it will be selling its "wares"?
- Is the product affordable in the targeted market segment?
- Can the product be produced and sold at an acceptable profit margin?

Many other issues should be addressed. These issues are generically addressed in Chapters 1 through 3. However, for the purpose of this chapter, the main point raised is that the marketplace does not care about technology. The marketplace cares whether or not the money spent on buying the product has been well spent. Consumers do not care how the product was made or that the product uses the most sophisticated hardware or software. The consumer only wants value for the money he/she has spent.

Figure 5-27 focuses on the services offered by WAP and iMode. Both support the following services:

- **E-Mail**
- **News** Current events, breaking news reports, weather, horoscopes, traffic, and so on.
- **Finance** News, online trading, stock watch, banking (at this moment iMode supports retail banking transactions), and so on.

E-mail
News
Finance News
Online Trading
Banking Transaction
Shopping
Screensavers
Games
Portal Access
Music
Others

WAP and iMode provide similar services.

FIGURE 5-27 Wireless Internet Services—iMode and WAP

- **Shopping** Retail, travel, musical concerts, entertainment shows, and so on. The wireless handset becomes a transaction processing point.
- **Messaging**
- **Screen savers** Like a "real computer," only iMode supports GIF images. Moving pictures are also supported by only iMode at this time.
- **Games**
- **Portals**
- **Entertainment** Online music, movies, online radio, and so on.

WAP and iMode support and present these services differently. However, despite the technical differences and the compatibility issues, the consumer seems enchanted with both product offerings. For now, the consumer is "playing with a brand new toy" in the wireless marketplace. The games offered by iMode appear to be more numerous, but it seems that this does not matter for now. Graphics capabilities of iMode are superior to WAP currently, but that should be only temporary because of WAP's evolution.

The WAP service providers are providing value to their existing customer base. WAP service providers are evolving their platforms so that they will be W3C (and therefore, WWW) compatible. This will enable WWW Internet sites and WAP to adapt their services to one another for cross marketing more easily. The iMode supporters already claim WWW compatibility. Cross marketing will enable both market segments to benefit.

At the time of this writing, it has been rumored that the iMode supporters are evolving their platforms to support WAP users by enabling WAP phones to access iMode content. This is being done in Japan today. It is one way for iMode manufacturers and service providers to sell more equipment and services. By enabling a WAP user to get iMode content, an iMode service provider could use the product as a way of convincing the WAP user to buy their primary service from the iMode carrier. More than likely, a gateway function will be used to act as a mediation and conversion access point. Figure 5-28 illustrates this concept.

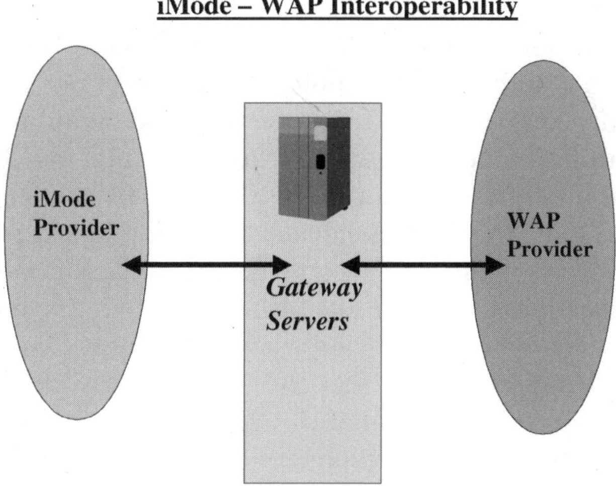

FIGURE 5-28 iMode—WAP interoperability

Converging technology has become a commonplace activity in the global economy. Convergence has become the business of many startups both in the software and hardware businesses as well as in the service provisioning businesses. As noted in *Telecommunications Internetworking: Delivering Services across the Networks*, convergence has become a necessity if one wants to survive and thrive in this business.

In its current form, WAP is limited by comparison to iMode. WAP does not have the breath of access to the WWW as iMode. One can even make an argument, albeit a poor one, that iMode is Japan-centric, but iMode is also being sold outside of Japan. However, this WAP situation seems temporary and barely noticeable to a user. He/she is too busy experiencing the variety of information services. The user understands that "change is inevitable" and he/she will get cost effective, usable, and useful product. Furthermore, the user usually does not care who provides it.

WAP VERSUS iMODE VERSUS NEXT GENERATION WIRELESS

To those in the industry, the issues of next generation wireless were originally conceived to be a packet data enhanced version of current wireless. Until WAP and iMode took hold as realizable applications, the next generation took the form of various debates on which radio technology was better. The investment community was looking for something to wrap technology efforts around and hang their flags on such as 3G (third Generation wireless supported by the CDMA carrier community) or EDGE (the TDMA carrier community's next generation wireless effort). WAP has become that flagpole for most people in the investment community. Although iMode is widespread, WAP appears to be more widespread at this time. One approach is a proprietary effort whereas the other has a wider breath of industry support. Note that at the time of this writing, a debate is occurring between the pioneers of WAP over patent rights.

One should not focus on intellectual property ownership. The proprietary nature of an approach is not an issue for many investment houses. In the end, it is the marketplace that will decide who wins. As trite as this may sound, it is the truth.

INTERNET CONVERGENCE

Convergence is a term that the telecommunications industry has used to describe the fading business and technical distinctions between the various telecommunications services segments. The Internet is causing the telecommunications industry to converge the various segments onto the provisioning of Internet services. In fact, the Internet industry must face that same convergence within itself: The internal convergence referred to within the Internet concerns WAP, iMode, and the WWW. The Internet and the users will benefit more from a wireless Internet that communicates (talks) with the wireline Internet than from a wireless Internet that is proprietary.

More than likely, XHTML will become the common markup language for both iMode and WAP. As noted previously, HTML is the universal language of the WWW. HTML is used to support communication between the user and the computer. XHTML is a combination of HTML and XML. The combined format will define the data and the presentation of the data. Common tools and standards (industry established and defacto) will serve as a lightning rod for convergence and growth.

One cannot overstress the need to ensure compatibility between WAP, iMode, and the WWW standards. The Internet industry has the opportunity to grow beyond all expectations. By creating synergies between the wireless Internet and the wireline Internet, the two industry segments can grow. The more customer utility/usefulness the Internet has that crosses market boundaries, the greater the value to the customer and hence, the larger the overall market. This scenario is akin to one in which a company spends money to grow the subscriber base in order to grow the revenue stream. Cross sales and marketing has beneficial effects on revenue.

SUMMARY

The Internet is a new information medium that is undergoing drastic changes within itself, as well as galvanizing an entire global economy. As the Internet technologies change, so will the economy. Until 1999, one could only dream of the Internet supporting voice. The wireline Internet was growing, but looking for a way of becoming the consumer's primary means of communicating. In 2000, it is moving toward reality because of IPv6. In 1999, the entire wireless community was looking at enhancing its networks; however, without any clear development goals, the correct evolutionary path was anyone's guess. In 2000, the excitement of the Internet galvanized the entire wireless industry toward a common set of goals.

CHAPTER 6

E-COMMERCE AND MOBILE E-COMMERCE

What is e-commerce? E-commerce is shorthand English for *electronic commerce*. It is a realization of doing business using Internet technologies. As noted in Chapter 2, "Challenges of the Internet Business—Business and Technology," e-Commerce is an activity/application and not a product to be sold to an Internet user. E-commerce has become an activity/application that has captured the interests of both the commerce and investment sides of the business community. E-commerce has been able to serve as a new home for existing ways of doing business. The new ways of doing business are translations of current ways of doing business and their associated relationships. These relationships are as follows:

Business-to-Business (B2B) The B2B relationship describes the relationship between two or more businesses selling services to one another (see Figure 6-1). A steel manufacturer buys coal from one company to use in smelters to turn iron ore into steel. The market has even created a sub-division of the B2B called *Business-to-Small Business* (B2SB) relationship.

Business-to-Consumer (B2C) The B2C relationship is the one the average consumer is familiar with: Retailers selling products online (see Figure 6-2). Consider this

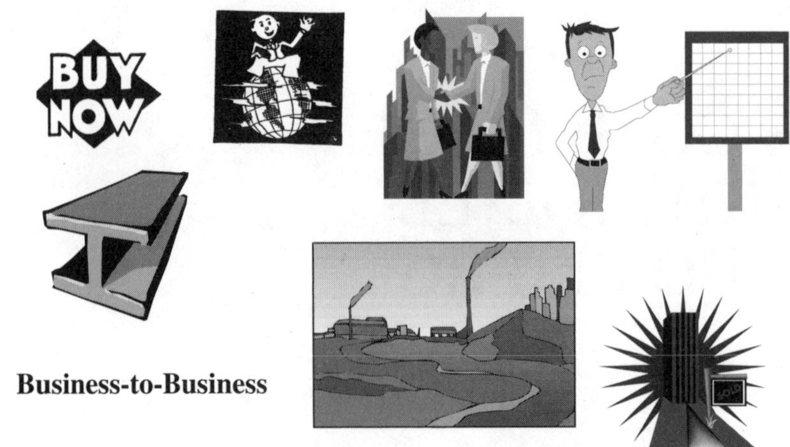

Businesses can sell one another products, raw materials, and services.

FIGURE 6-1 B2B

FIGURE 6-2 B2C

online catalogue shopping for clothing, books, vehicles, and so on.

Business-to-Government(B2G) The B2G relationship is a new one that is growing over the Internet (see Figure 6-3). This relationship is about the business community doing business with sectors of the government. Sectors of the government have been able to use the Internet to purchase supplies and broadcast industry-wide bulletins. The government, whether local, state, national, or international, is the single largest customer of goods and services. To many people in the telecommunications business, the government is the single largest customer in the marketplace.

Exchange-to-Exchange (X2X) The X2X relationship is the linkage of multiple B2B relationships/ marketplaces (see Figure 6-4). X2X e-commerce is the next logical step beyond B2B or B2SB. Companies are able to purchase supplies/products at competitive prices. X2X can be likened to a farmers' market; where large food distribution companies bid for farmers' products. Manufacturers and vendors of all kinds seek to purchase supplies and services over the Internet. X2X e-commerce could ultimately result in lower prices to the consumer because of the lower costs to the retail outlets and to the distributors (that is, the middleman).

Although it may sound facetious, another relationship exists called the *Taxpayer-to-Government* (T2G) relationship (see Figure 6-5). Today, taxpayers can file their income taxes over the Internet. They can find a variety of information from the various government databases. Furthermore, taxpayers can file any number of requests for information (from government agency) over the Internet. For example, during 2000, a global vehicle tire recall was put into effect. The various national governments involved communicated with the public via the news

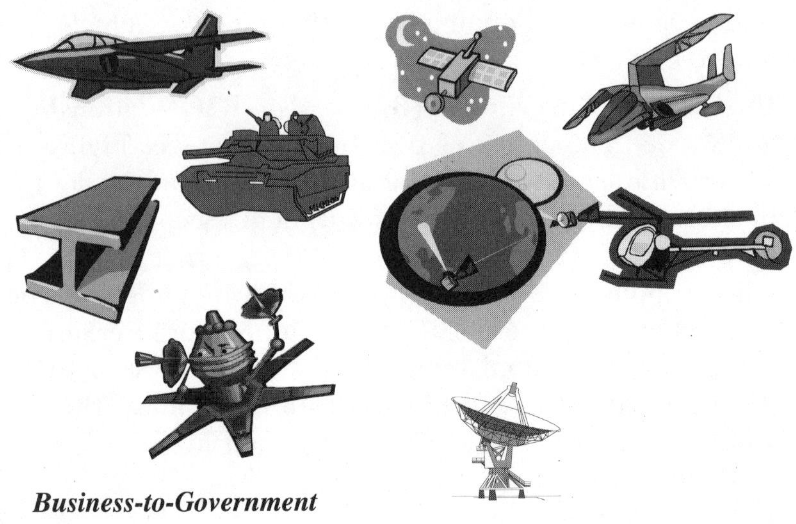

FIGURE 6-3 B2G

FIGURE 6-4 X2X

Paying taxes

Filing complaints

Speaking to the local representative

Responding to polls

Citizen communicating with government

Taxpayer-to-Government

FIGURE 6-5 T2G

media and the Internet. Many individuals filed complaints to the various national transportation safety authorities via the Internet.

These relationships exist today in the non-Internet business world; however, the Internet has been able to assist business people in reaching a much broader audience faster than before. The aforementioned relationships are finding their way into the wireless marketplace. These relationships form the basis of m-commerce.

WHAT IS M-COMMERCE?

The changes that are occurring in the Internet have been occurring on several business and technical fronts, which are as follows:

- New and improved *Internet protocol* (IP) called IP Version 6.
- Overlaying the IP over existing network signaling protocols such as SS7 and ATM.

- Faster and bigger routers.
- Faster network access.
- Creating business alliances that result in merging new types of content in the Internet business.
- New companies focused on serving specific pieces of the Internet business, such as outsourcing operations, Web site creation, Web site administration, and e-mail.
- Improved Internet encoding formats and scripting languages.
- Wireless Internet technologies.
- New types of terminal devices such as mobile handsets, Internet-only desktop and laptop computers, small, palm-sized terminals, and so on.

The changes are occurring at an eye-blurring, neck-breaking pace, yet the most daunting effort is finding a way to bring the Internet to the mobile user. The mobile (wireless) marketplace has been growing at an explosive rate. The wireless marketplace is highly competitive and expanding. Today, wireless is seeking to replace the wireline telephone via wireless local loop technologies. The Internet has added a whole new dimension to the wireless market. The Internet community has even coined a new term to identify wireless' place in e-commerce. The term m-commerce is short for mobile e-commerce. Figure 6-6 depicts the mobile e-commerce marketplace.

M-commerce brings wired e-commerce to the user via a wireless handset. M-commerce is just being introduced now. However, it faces some challenges. To most users, m-commerce is simply a new way of shopping for durable and non-durable goods. This is not a trivial application. A nation's economic health is partially based on the amount of retail products sold. However, m-commerce can bring far more to the marketplace. The challenges facing the mobile e-commerce marketplace are both technical and business-related.

FIGURE 6-6 M-commerce

M-COMMERCE ISSUES

M-commerce is an application that has the potential of bringing the Internet and wireless telecommunications to the average consumer. However, consumers that use the wired Internet for shopping are currently able to:

- See pictures of the product they are purchasing.
- Print copies of the Web pages they are viewing.
- View multiple windows showing products.
- Have easy to view instructions on purchasing.
- Have colored pictures.
- View consumer product information, that is, the label information most consumers read on the back of a box.

The previous list should not be discounted. Consumers only buy when they can see the product. Recent retail shopping disasters over the Internet have caused many to shy away from Web site shopping. However, these disasters have not caused a decline in shopping. Rather, the consumer is even more knowledgeable now than before. Consumers want to see the product they wish to purchase, especially if they cannot touch the product. The challenge for the wireless community is that mobile handsets have limited screen real estate. The displays are typically no bigger than one inch by two inches. One can make a bigger display, but that would result in either a larger handset or less space for the physical keypad. Neither option is acceptable. The question remains: How do you give the wireless Internet user the same kinds of service that the wired Internet offers?

The wireless carriers are pursuing two different wireless Internet protocols. One is a stripped down version of IPv4 and other WWW technologies, whereas the other is totally incompatible with the existing WWW. In order to expand the wireless Internet, full access to the WWW is essential. The entire Internet customer base surfs the WWW. Without interoperability between the two Internet types, the wireless Internet will suffer. Before going any further, it is necessary to discuss mobility.

What Is Mobility?

Mobility has certain attributes that create challenges for Internet Web browsing. It is important to understand mobility before delving any deeper into the subject of m-commerce. Mobility can be described in a number of ways. One could call cordless telephones mobile devices because they are wireless. On the other hand, a cordless user cannot leave the vicinity of their home (that is, the cordless base station), which limits range and therefore, possibly limits the device's utility. Mobility has the following attributes:

- **Call delivery to a mobile subscriber** (home and visiting in another network).

- **Roaming** (supporting calls to and from visiting subscribers).
- **Authentication** Authentication is the process of user identity confirmation. Identity confirmation can involve checking handset/terminal device identity by interpreting "secret" keys/data messages. If the data keys/data messages have been altered or do not show a specific format, the call will not be completed.
- **Validation** Validation is often confused with authentication. Authentication essentially certifies the user as a either "real" or "fake" and either "good" or "bad." Validation certifies the "permission" to complete the call. For example, a user calls another party using a mobile handset. The carrier certifies that the handset is the real one and is not a clone being used by an unauthorized person. After the handset has been given the "thumbs up," the carrier checks if the call is allowed under the user's billing plan. This next step is validation (the final "green light" to complete the call).
- **Handoff** Handoff is the process of reassigning subscriber handsets to specific radio channels as the handsets move from cell site to cell site.
- **Path minimization** (efficient routing of a mobile call) Path minimization is the process of the efficient fixed network (the non-wireless portion of the call) routing of wireless call tables in the wireless carrier switches.

Figures 6-7 and 6-8 illustrate the concept of mobility. The reader should understand that unlike wired computers, mobile handsets should be viewed as moving IP addresses. More information can be found on wireless in *Telecommunications Internetworking: Delivering Services across the Networks*.

Mobility comes in two types: personal mobility and terminal mobility. Personal mobility is defined as the subscriber's (user's) ability to make and receive calls regardless of location and the type of terminal equipment the call is being made from or to.

Personal mobility supports independence from specific terminal devices and locations. At this time, most users of

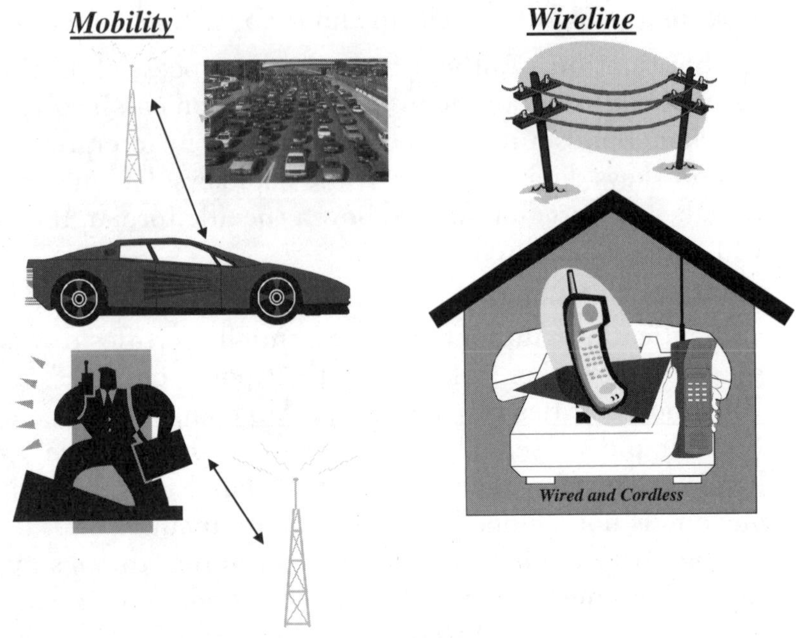

FIGURE 6-7 Mobility

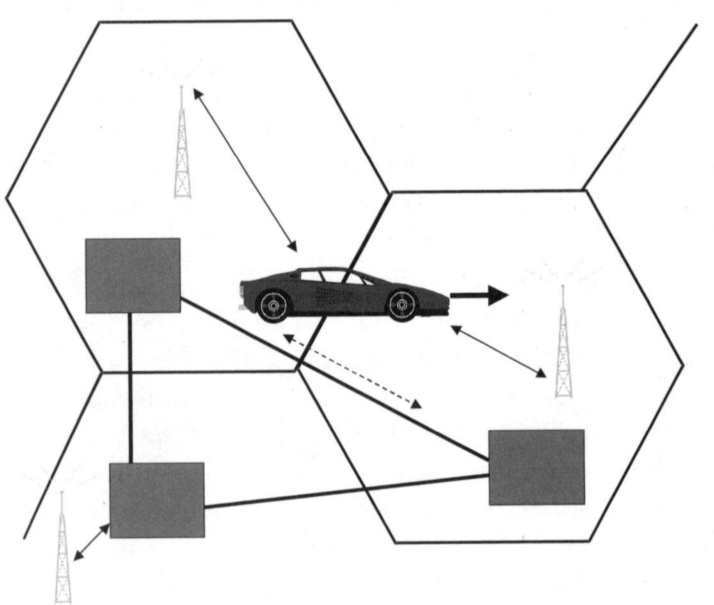

FIGURE 6-8 Mobility: Handoff and path minimization

telecommunications services are tied to a specific handset or fixed termination point (for example, termination at a home or business). Personal mobility requires the use of personal numbers or some other form of personal identification (for example, smart cards) to identify the user.

Terminal mobility is defined as the subscriber's (user's) ability to make and receive calls regardless of location. Terminal mobility infers that the network will authenticate and validate the terminal and not the person (user).

Figure 6-9 illustrates the concept of personal and terminal mobility.

Mobility adds a unique dimension to the Internet. The Internet must somehow find the user. The user can be anywhere. The visited network needs to communicate the user's location to the home system so that the home system knows how to route calls to the user. In the case of the wireless Internet, the network needs to be able to find the user as well. The IP has no parameters to support this attribute. WAP and i-mode are designed to support mobility. IPv4 and IPv6 do not support mobility. In order for WWW IP networks to communi-

<u>Personal Mobility</u> - Personal numbers or ID numbers. Calls delivered to the person.

<u>Terminal Mobility</u> - Calls directed to devices with unique numbers. Calls delivered to the device.

FIGURE 6-9 Personal mobility versus terminal mobility

cate with WAP or i-mode networks, the mobile characteristics of the user must be accounted for. In all likelihood, the mobility information will be managed by the wireless carrier's network. Operating basics of a wireless carrier can be found in my book, *Telecommunications Internetworking: Delivering Services across the Networks*.

M-COMMERCE TECHNICAL ISSUES

The following list includes some of the technical issues surrounding m-commerce that have been identified up to this point:

- Handset display screen's limited small size.
- Mono-color display screens.
- M-commerce "wannabees" should be educated on the nuances of mobility.
- User interface—Voice commands and prompts are supported. Human factors engineering will find this a challenge because if wireless Internet is going to rise above and beyond movie listings, concierge services, and stock reports, an easier way of browsing the Web must be found.

These technical issues concern form, not signaling. Remember that the user's focus is usability and not the nuances of technology. These issues are not minor issues, but are in fact potential "show stoppers" if they are not properly addressed. See Figure 6-10 for an illustration of the aforementioned points.

The form factor issue has two views. One view, called the "anti-wireless Internet" view, is that the handset is used for direct communication with other people. Furthermore, voice navigation through a series of menus is difficult on a regular wireline telephone; making a wireless user do it for Web-based services will be no different. This view also purports that the desktop and laptop computers lend themselves to the "network effect" of the Internet. In other words, one can view multiple

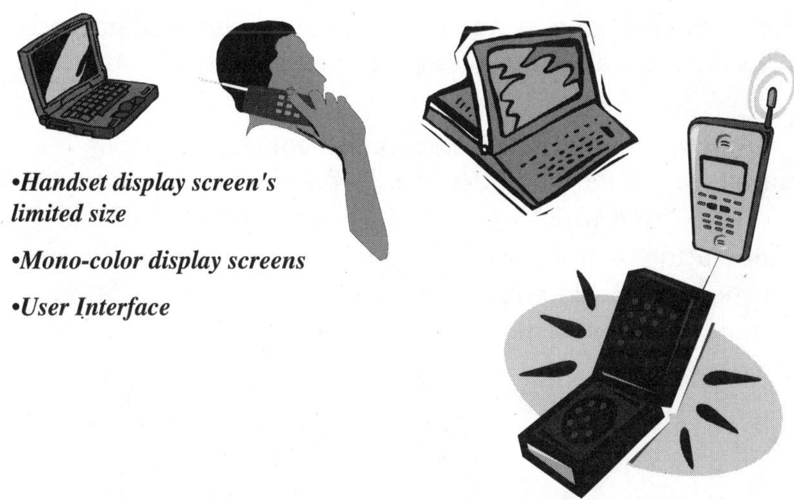
•Handset display screen's limited size
•Mono-color display screens
•User Interface

FIGURE 6-10 M-Commerce—Mobile issues

screens while on the Internet. One can also transmit a URL to multiple people easily. Note that these same people also remind users that the reason why the computer industry is working so hard to expand the viewing real estate (viewscreen size) of the display screen is because people want to see as much as they can.

The "anti-wireless Internet" view also believes that pagers provide the kind of stock quote information, news, and text messaging services that the wireless Internet supporters want to provide.

The other view, called the "pro-wireless Internet" view, supports the development of Internet access via the wireless handset. This group of supporters, which is everyone in wireless, is conducting a variety of market and technical research to find the best way of packaging this product for the consumer. From a technical perspective, the handset manufacturers and carriers are conducting various research to determine what the wireless Internet users want and how to present the information to them.

Applications service providers (ASPs) and content providers currently work in the wireline Internet and are attempting to find a place in the mobile marketplace. The challenges these ASPs and content providers face are as follows:

- Understanding the wireless technical challenges of mobility, which includes handoff and roaming. Many ASPs and content providers lack the knowledge of the basic operating characteristics of a wireless system. How can one expect to provide the most basic support if one cannot understand how the customer's network works?
- Modifying existing software and hardware to support mobility.

The previous list is not an exhaustive one and it should not be at this time. M-commerce is so new that all of the wireline Internet vendors are rushing into the mobile space without understanding the fundamentals of wireless communications technology. Figure 6-11 is a rendering of the need to understand the fundamentals.

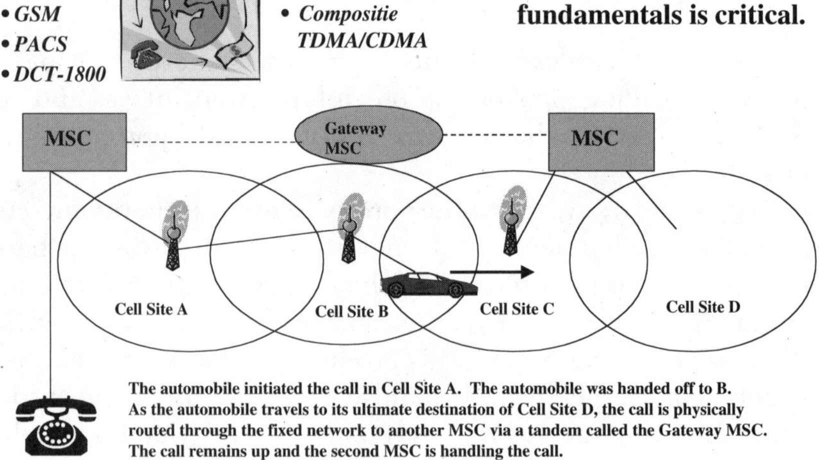
• TDMA
• CDMA
• GSM
• PACS
• DCT-1800
• DECT
• Wideband CDMA
• Compositie TDMA/CDMA
Understanding the fundamentals is critical.
MSC
Gateway MSC
MSC
Cell Site A
Cell Site B
Cell Site C
Cell Site D
The automobile initiated the call in Cell Site A. The automobile was handed off to B.
As the automobile travels to its ultimate destination of Cell Site D, the call is physically routed through the fixed network to another MSC via a tandem called the Gateway MSC.
The call remains up and the second MSC is handling the call.

FIGURE 6-11 The wireless fundamentals

M-Commerce Business Issues

The business issues surrounding m-commerce are far more difficult to grasp than the technical issues. Before a service or product is introduced into the marketplace, market trials are conducted. Mobile-based Internet services are being sold today. However, the industry has yet to understand the extent of the market opportunities. The din of noise created by the rush into the mobile business space has been caused by thousands of companies claiming they are working in the m-commerce space. Many people in the investment community are unsure of the opportunity, but it is understood that wireless is growing. It seems logical to assume that if wireless is growing, so will every other opportunity that is associated with it. However, this is a poor assumption to make.

As noted in Chapter 3, "Challenges of the Internet Business—Business and Technology," the danger of doing business over the Internet is that it is impersonal and people like speaking with other people. Questions should be asked such as: Does your existing wireless customer want to access the Internet? If the answer is yes, then what does the customer want from Internet access?

These questions are not trivial. They go to the heart of whether or not one should pursue m-commerce. Market research also needs to be conducted. The Internet space is so crowded on the wireline Internet side of the business that the wireless carriers' only difference is mobility. Mobility is a major difference, but Internet users have become accustomed to surfing high-bandwidth multimedia Web sites. What kind of Web access will the wirelss carriers provide?

The value chain analysis that the wireless carriers must conduct must account for the current players in the Internet space. ASPs and content providers need to conduct the same value chain analysis. As noted in Chapter 3, a value chain is a business tool that is used to describe the value of a sequence of events or relationships. A value chain analysis is just one of the many market analyst tools that should be employed. See Chapter 3 for additional information on general marketing issues. The m-commerce space is new, and time is limited for a business analysis.

The following are basic questions that should be raised when entering the m-commerce business space:

- What is the product?
- Who is your customer? This would include demographics, buying behavior, decision-making process, and so on.
- Is the product easy to use?
- Does the product fill an existing need?
- Does the product make a task easier to perform?
- Does the product enhance the way a task is performed?
- How will you deliver the product or service to the user?
- Does a market large enough to warrant making the product exist?

These are basic questions that many people in the mobile business may be ignoring in their rush to provide wireless Internet access. Figure 6-12 is an illustration of these basic questions.

Basic Questions	**Who?**
What is the product?	**What?**
What does the product do?	**When?**
Who is the customer?	**Where?**
How is the product delivered?	**Why?**
How is the service delivered?	**How?**

FIGURE 6-12 Asking basic questions

Scanning through the various trade journals, one will find various Internet accessible products being deployed while the WAP standard and the i-mode specification are under development and enhancement. One should ask the question: Will this equipment be useable in a year from now?

The potential danger of rushing into any technology-based business endeavor is incompatibility between equipment and standards. For those whose telecommunications career began in the world of the Internet, this occurrence may sound impossible. These same people may claim that a particular company's market dominance will dictate which standard will be followed. Such claims may be true; however, similar thoughts have occurred in other high-stake, technology-business gambits. Figure 6-13 depicts the question of selling products ahead of the market.

THE DANGER OF MOVING WITHOUT PROPER PLANNING

An example of rushing ahead without proper industry-wide planning concerns the service *Integrated Services Digital Network* (ISDN). ISDN development began in 1976 and was not completed until the mid-1990s with the deployment of the Regional Bell Operating Companies' National ISDN-1. Until that moment, small and scattered deployments were available. The problems faced by ISDN proponents concerned timing.

SWOT

Thinking through the operational issues

Thinking through the market issues

Market dominance is a goal

Size doesn't matter

Planning

Execution

Good Business Plan

FIGURE 6-13 Rushing ahead

ISDN aimed to provide the subscriber voice, video, and data on a single pair of wires. In its early days, ISDN was radical and would be the first protocol to introduce out-of-band signaling in the wireline local loop and the first to bring this set of services to the home and business user. The tragic state of affairs that took place in the mid-1980s was that carriers were selling ISDN services before terminal equipment had been developed and sold. ISDN terminals had been developed, but no terminal equipment was available that had standard interfaces for the user or network interconnect. This situation did not prevent vendors and carriers from selling ISDN services and terminals. Ultimately, the situation created a problem with the customer in the areas of customer care, repair, installation, and training.

Even though ISDN was meant to function as an end-to-end service (people to people), it could only do so within a *local exchange carrier's* (LEC's) central office. The LEC network at that time could not support out-of-band signaling interswitch or intranetwork. Therefore, without a *Signaling System* 7 (SS7) network, the full capabilities of ISDN could not be realized. The vision of ISDN overwhelmed the technical, business, and market realities, and created mayhem in the development of the market and technology. ISDN and SS7 were meant to function together. Unfortunately, they were treated as separate LEC planning efforts and deployment efforts; the disconnection in planning and deploying hurt the roll outs of both technologies. During the 1980s, ISDN was considered the subscriber interface to access the power of the SS7 network. Ultimately, because only a few large equipment manufacturers could provide SS7, the industry standardization discussions were far more orderly to manage. However, in the case of ISDN, dozens of small, mid-size, and large manufacturers providing ISDN terminal equipment were available. This is similar to the current state of the m-commerce and wireless Internet environment.

So much time and money had been invested in ISDN terminal devices in the manufacturing community and the subscriber base that bringing the industry together was a nearly impossible task. Eventually, the industry found a way to work together. However, the two victims in the 20-year industry effort were the technology and the customer.

Part of the problem with the early ISDN marketing efforts was the technology focus. Subscribers do not care how the service is provided; they simply care if the services have value and work consistently. Unfortunately, by the time all of the ISDN marketing efforts took flight, the industry and the subscriber base had new technology alternatives that provided them with voice, video, and data. The danger being faced by those rushing to enter the m-commerce business space is the same danger suffered by ISDN. Even closer to "home" is the recent spate of Internet-based company bankruptcies. M-commerce is another business opportunity that has the potential of losing "steam" unless careful planning is done to understand the market and to meet the market needs. Figure 6-14 is a rendering of the need to plan properly.

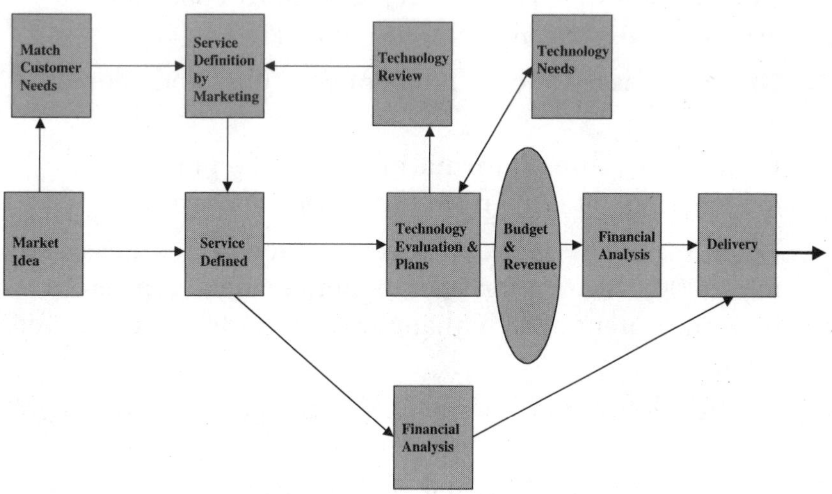

FIGURE 6-14 Proper industry planning

M-COMMERCE SERVICES

The m-commerce marketplace is new. M-commerce is a mobile version of the e-commerce marketplace and both are in their infancy. If one picks up a wireless handset and wanders through its Web services, he/she will find the following available to the user (see Figure 6-15 for a depiction of m-commerce services):

- **Financial news** General business information that is updated on a periodic basis.
- **Personalized stock portfolio watch** The user identifies specific stocks about which the m-commerce site is to transmit stock quote information.
- **News media reports** Current events. Often broken down into the categories of national and international news.
- **Online stock trading** Trading on the stock market via specific online trading companies.
- **Banking transactions** Depositing money or transferring money between accounts.
- **Entertainment** Ticket purchases, horoscopes, movie listings, lottery results, and movie reviews.
- **Directory assistance** Telephone number and address display.
- **Retail store sales information** Based on location information obtained from the carrier's network.
- **Location of the nearest store or place of business** As of mid-2000, the location information element appears to have the potential for enhancing the wireless Internet for the user.
- **Auction locations** Some people consider auctions a great place to shop.
- **Traffic reports** Vehicular road conditions and airport status.
- **Weather reports.**

- **Internet search engines** Via business relationships, specific Web search engine companies agree to create abbreviated versions of their Web pages for the sole purpose of searching the Internet.
- **E-commerce Web sites** Those that are normally accessed via the wireline Internet.
- **Portal access** Many wireless carriers have created their own portals to specific "hot types of sites."
- **E-mail.**

The question that has arisen with many wireless carriers providing Web access is: Are you providing the same broad level of access to various sites as the wireline WWW? The answer is no; however, most people who are currently buying the service do not care because expectations at this stage of m-commerce growth are low and wireless users do not know what they want.

FIGURE 6-15 M-commerce services

Currently, the accessible Web sites are restricted to limited number of sites via a contractual relationship with the wireless carrier. Without total compatibility with the WWW, the various versions of wireless Internet and the associated m-commerce applications will be restricted to a limited number of sites. This may actually be a blessing in disguise because so many Web surfers are offended by the poor quality of the search engines that a carrier willing to filter out the unpopular sites may be perceived as performing a useful service. See Chapters 2, "Business Models" and 3, "Challenges of the Internet Business—Business and Technology," for additional information on search engines and portals.

As noted in Chapters 2 and 3, the current basic and generic ISP services include the following:

- E-mail
- File storage
- Web site creation
- WWW access
- Search engine
- Chat rooms
- Internet portal access
- *Voice over IP* (Voice over Internet Protocol)

The aforementioned services may be offered by the wireless carrier, but the question remains of whether or not this is necessary. E-commerce is a subset of the aforementioned listed ISP services. Furthermore, all of the services listed as being provisioned by the wireless handset are also provided within the structure of services listed as residing on the wireline Internet platforms.

The services listed earlier in this section are accessible over Web-enabled handsets. However, many of these handsets have no graphics capability and only support text. These handsets do not support audio or video streaming at high-bandwidth rates. This raises the following questions:

- Does the wireless user want to have the same kind of Internet services that he/she has on his/her wireline Internet computer terminal, also on his/her mobile handsets? This question is being answered in the marketplace at this time.
- Is m-commerce nothing more than mobile wireline e-commerce or is it something more or different?

Figure 6-16 illustrates the aforementioned questions.

MOBILITY AND M-COMMERCE

Rather than thinking about how to duplicate what the wireline Internet provides now, the wireless industry should think about how to further enhance the wireless carrier's greatest asset:

Mobile e-commerce does not have to look or act like wired e-commerce.
•Wireless systems and wireline systems have different attributes.
•Wired e-commerce serves customers that are stationary.
•Wireless (mobile) e-commerce serves customers that are moving.
•A basic and inherent difference exists between the customer segments.

FIGURE 6-16 Does m-commerce need to look like wireline e-commerce?

mobility. The wireless industry raises the following questions concerning mobility:

- What kind of information does a mobile user want to see on a moment's notice?
- How much time does a mobile user want to spend punching in information one keystroke at a time?
- How much useful and useable information can be displayed on a small display?
- Can the mobile user access the wireless Internet via another kind of terminal device?
- Does the mobile user really care about receiving local retail sales notices like one receives news media pages on his/her wireless devices? Local sales notices will be presented based on the user's location.
- Does a mobile user buy a product over the Internet without seeing a picture of it?
- Is the mobile user willing to use his/her handset as the principle means of accessing the Web?
- Does the mobile user have any desire to trade stock over the wireless Internet?
- Because people like to keep a hard copy of transactions, can the mobile user print his/her transactions on a printer?

Figure 6-17 depicts the need to focus on wireless' strength.

It is a difficult sell to convince a current WWW surfer to suddenly translate his/her Web surfing habits from a 19-inch display screen into surfing on a display screen of one inch to 2.5 inches. In order for the wireless carriers to expand their Internet opportunity, they must exploit wireless service's greatest attribute: mobility. In other words, rather than simply providing wireline services that can be used over a wireless handset, the carriers ought to find a way to provide an Internet service that is either dependent on or directly related to mobility. The next section will provide a brief explanation of this view.

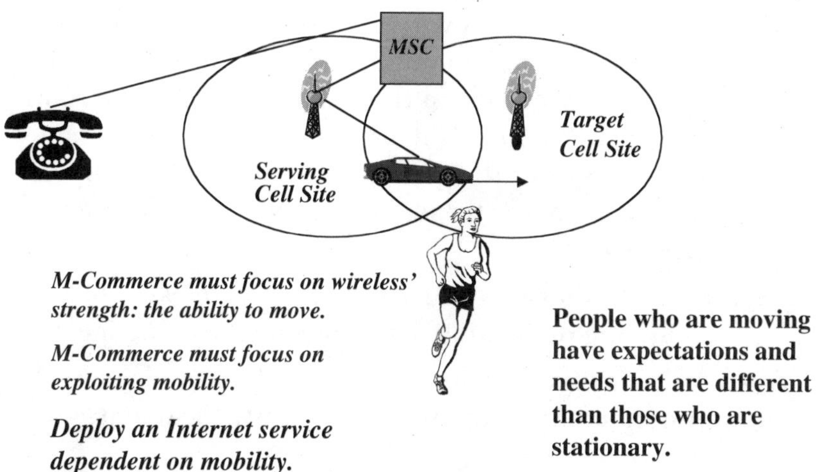

MSC
Target
Cell Site
Serving
Cell Site
M-Commerce must focus on wireless' strength: the ability to move.
M-Commerce must focus on exploiting mobility.
Deploy an Internet service dependent on mobility.
People who are moving have expectations and needs that are different than those who are stationary.

FIGURE 6-17 Mobility: Asset and difference

MOBILITY LOCATION INFORMATION: AN ASSET

Since 1994, the wireless industry has been working to deploy location technology in order to meet a United States *Federal Communications Commission* (FCC) mandate for cellular and *Personal Communications Service* (PCS) carriers to implement location technology for the purposes of determining the geographic location of 9-1-1 calls. The primary purpose of the equipment is to support public safety; however, the location technology can be used to support location-based services. Both the mobile and Internet industries came upon the idea of using geographic location information to enhance the value proposition of wireless Internet.

Geographic location is an information element that the network can use to trigger an action within the network. This thought process was the basis for intelligent networking in the wireline telecommunications world. The wireless Internet can also use location to act as a trigger for the network to perform a specific action supporting a specific service. The following types of services can use location information:

Directed advertising A wireless user drives or walks into an area that has four shopping malls. The wireless provider registers his/her presence in the area, but more importantly, determines that the user is near a specific shopping mall. The wireless provider then issues a wireless sales coupon to the user over his/her Internet device for a store located in one of the malls.

Enhanced concierge 4-1-1 service A wireless caller requests driving or walking directions to a specific location.

Traffic directions A wireless user is instructed via his/her Internet device on a route to avoid a traffic accident.

Figure 6-18 is an illustration of how location information can be used to support the wireless Internet.

The previous list is a small sample of what wireless Internet pundits are considering for deployment in the mobile Internet space. Location information is a necessary element in order to make any type of wireless call setup or data transaction. The wireless network, including the wireless Internet, must know

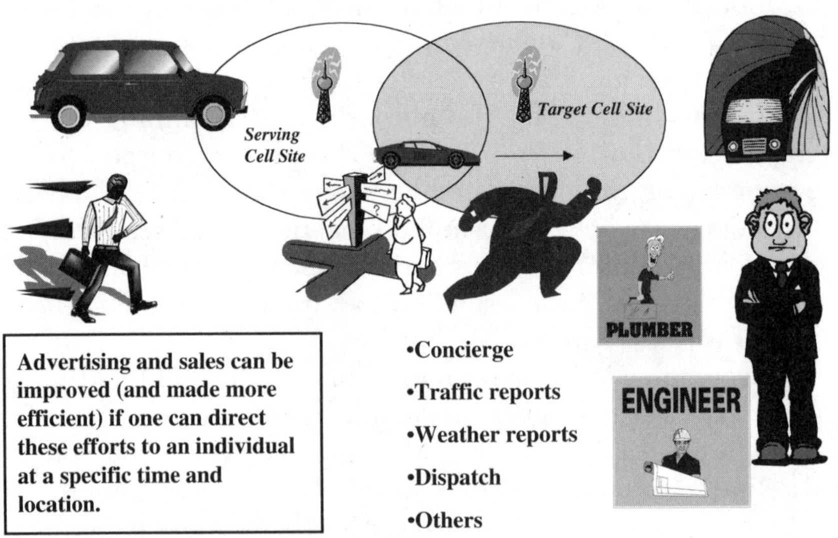

FIGURE 6-18 Location information and the wireless Internet

where a person is in order to ensure that the call or datagram arrives at the right location. The location technology being deployed for public safety purposes provides the network with a granularity of location information (and in a format) that it never had access to before. The mobile Internet business space has been seeking some way to differentiate itself from the wireline Internet to entice mobile users to buy Internet access from the wireless provider. The mobile Internet community has come to understand that they should exploit the biggest difference between wireline and wireless user communities; one is mobile and one is not. Location technology is the key to mining the value out of the mobility attribute and ultimately creating value for the wireless Internet.

SUMMARY

The obvious benefits of purchasing over the wireless handset are:

- No more waiting in check-out lines.
- Real-time checking of product sales.
- Shipping product to the home.
- The handset is a highly transportable device.

M-commerce is still in its infancy. To many people, m-commerce is a boondoggle. To others, m-commerce is a positive economic force. I believe that the m-commerce market will take off when the carrier community changes the view of what a mobile handset is. In other words, PDA-like devices have a place in the world of m-commerce. Also, the display screen should be bigger in order to display a larger breath of information. However, the information displayed does not have to be an exact replica of what is seen on the wireline Internet Web sites. The fact is many of the Web sites today are so cluttered with advertisements that viewing them is painful to the eye. A small display screen may be an asset for the millions of users who find

- *Small display screens*
- *Replicating the wired Internet*
- *Does the wired Internet have to be completely duplicated?*
- *Incompatible standards*
- *Identifying services*
- *Identifying markets*
- *Is the wireless Internet competing with the wired Internet?*
- *Can wired e-commerce co-exist with m-commerce?*
- *Is there a synergy between wired e-commerce and m-commerce?*

FIGURE 6-19 M-commerce issues

banner advertising a major annoyance. Thinking out-of-the box will be a necessity. M-commerce players must do what they can do avoid trapping themselves within the operating paradigm of the wireline Internet; a problem for one is not necessarily a problem for the other industry segment. Figure 6-19 is a depiction summarizing all of the issues surrounding m-commerce.

Finding the correct services will be a challenge in an environment that is not necessarily giving the industry time to conduct proper research. The market research conducted today is primarily empirical: gathering data while selling the service.

CHAPTER 7

CUSTOMER CARE, BILLING, AND REVENUE ASSURANCE

Customer care and billing are the least understood areas of business in the Internet space. Both wireless and wireline Internet communities have been rushing into the marketplace with brand new services without a clear understanding of how to handle customer complaint issues or how to bill the user. This does not mean that customer care and billing are not important. Customer care and billing are not "sexy" technology issues to write or even talk about. Customer care and billing are not the core services of the various Internet companies in the marketplace. However, without customer care and billing system processes in place, these Internet companies will go out of business. As noted earlier in this book, the inability to execute business plans has been as much, if not more to blame for Wall Street's wariness of the Internet than the lack of ideas. This inability to properly execute extends to the lack of proper customer care and billing systems. Figure 7-1 is a depiction of the perceived role of customer care and billing.

Revenue assurance is new to the telecommunications industry, having been aggressively introduced in the late 1980s. No single revenue assurance process or platform exists. This will be discussed later in this chapter.

YOUR

Customers expect to be treated courteously.
The customer relationship can be shattered easily and quickly due to a misspoken word.
The customer pays the bills.
Customer care and billing systems are usually the last things considered by a provider.

FIGURE 7-1 Customer care and billing: Perceptions

CUSTOMER CARE: WHAT IS IT?

Customer care is simply what the title implies; it refers to every aspect of caring for the customer's questions and needs. Customer care is a process that links the customer directly to the core functions of the service provider. To properly understand customer care, one should immediately stop thinking that the customer understands, let alone even cares about how a company's technology or equipment platforms function. Much of the hype in the investment community is about the technology and the ability to turn intellectual property into a profitable business opportunity. The problem with the various sales pitches in Wall Street and elsewhere in the financial community is about technology, not about what it takes to bring the technology to life. In the case of service providers, the sales pitch is about technology and how it can differentiate the provider from other providers. However, these carriers fail to realize that if one's service looks like another carrier's wireless technology, the user perceives it as being the same. This is the same type of issue that PCS faced (unsuccessfully) when it was compared to cellular. Everyone in the wireless community talked about how different cellular and PCS technology were. The tragedy was many investors wasted so much time trying to understand that PCS was another wireless opportunity in another frequency band that precious time was lost in properly assessing the various PCS opportunities. Figure 7-2 highlights the importance of focusing less on the idea and more on the need to maintain the customer.

M-commerce and e-commerce have taken over the current headlines. However, it is likely that the multitude of failures in the areas of m-commerce and e-commerce will soon appear. Some of the reasons why these commerce sites will fail include a weak business plan and poor execution. However, the trade press will not discuss the poor customer care and billing systems. This is not only because customer care and billing are media "un-sexy," but also because something so basic is embarrassingly overlooked.

Providing service is not about the best technology, it is about serving the customer.

The customer relationship must be priority one.

FIGURE 7-2 Focus on the customer

Customer care processes address the following issues and questions:

- Customer orders for service—The provisioning of service for customers requires a structured framework to process those orders.
- Customer service order fulfillment process—A customer must be able to easily place an order for services. The harder it is for the customer to obtain what they want, the easier it becomes for the customer to decide not to do business with the service provider.
- Customer complaint reception process—Who does the customer call when a problem arises?
- Customer complaint resolution process—How a complaint is generally resolved must be clear to the customer at the beginning of the customer's call for new service or repair. Customers must feel that they will not be shuffled off from one menu to another or one person to another. This is linked directly to repair services for the customer.
- Methodology for addressing customer questions over the customer bill for services.

Figure 7-3 illustrates the customer care process.

•Customer order for service
•Service order fulfillment process
•Complaint desk
•Complaint/Trouble resolution process
•Billing Q&A
Know the customer.
Communicate constantly with the customer.
"Speak with" and "not to" the customer.
Respect the customer, because they pay one's salary.

FIGURE 7-3 Customer care process

WHAT COMPRISES CUSTOMER CARE?

The aforementioned must be supported internally and externally. The external view is what the customer sees. The internal view is what the customer cannot be allowed to see. The general rule of thumb is that the customer really does not care about "how one gets it done" just that "it is done."

External View

Customers want or expect a few basic things from a service provider:

- Questions answered quickly.
- Questions answered with as few carrier/ISP employees as possible.
- Issues resolved in a timeframe that is perceived as reasonable to the customer.
- The customer must have a high level of confidence that their questions will be professionally addressed. Therefore, the customer care professional's presentation is extremely important.

Generally, customers do not want to hear any "lip service," "sass," or "attitude" from the service provider. The customer care employee/professional is the face of the ISP, ASP, or other type of Internet company. Professionalism is a must in all aspects of dealing with customers and is essential in all business dealings. Figure 7-4 depicts customer expectations from customer care.

As noted previously in this book, ISPs divide customer care into the two following categories. The ISP maintains these external functions as the face of his/her company.

- **Customer care/complaint** The ISP interacts with the customer on the following matters: new service activation,

- **•Answer questions quickly.**
- **•Be professional and courteous.**
- **•Tolerate the screaming customer.**
- **•Customers do not like having to be passed on to other people; therefore, answer the question.**
- **•Remember the customer satisfaction is Job One.**

FIGURE 7-4 Customer expectations

billing inquiries, billing complaints, access telephone numbers, rating plans, and purchased feature capabilities.

- **Technical support** The ISP often maintains a separate "hotline" for technical questions. These technical questions can concern software installation, software troubles, access telephone numbers, and almost any question that the ISP cannot classify as a non-technical question. Often, the ISP will request that all questions be submitted to him/her via e-mail. This is not the most timely way of handling customer complaints because it can take 24 hours to respond to the customer.

The need for personal interaction cannot be overstated. A technical aspect of customer care also exists. Customer care systems are not single pieces of equipment, but are tightly coupled with an array of other systems. These other systems are as follows:

- **Service order provisioning systems** Provisioning of new service.
- **Accounts payable** Customer account status.
- **Accounts receivable** Customer account status.
- **Trouble ticket reporting systems** Customer reporting of troubles. However, the system also reflects system troubles that are not necessarily reported by customers, but instead are reported by employees.
- **System recent change** Customer care professionals are kept apprised of system changes and updates. In the event a customer calls and wants to report a trouble, the customer care professional may be able to explain that the trouble is only a temporary situation due to system changes.
- **Customer recent change** This system reflects customer changes to his/her own service profile.
- **Customer service profile** This system reflects the customer's current service profile.
- **System flashes** The system flashes are a reporting tool of activity within a provider's core switching/system infrastructure.

The aforementioned systems are integrated into a single interface platform for the customer care professional. Figure 7-5 is a rendering of the customer care systems. This platform enables the customer professional to input information that another organization must act on. For example, a customer calls for a new function called "unlimited online time per month." This new function or service is a simple billing rate plan change, which must be reflected in the customer service profile, which in turn automatically generates a customer recent change, and finally sends a message to accounting (payable and receivable). All of these events are automated. The customer will note that they spoke to a customer care professional, which is all the customer should see.

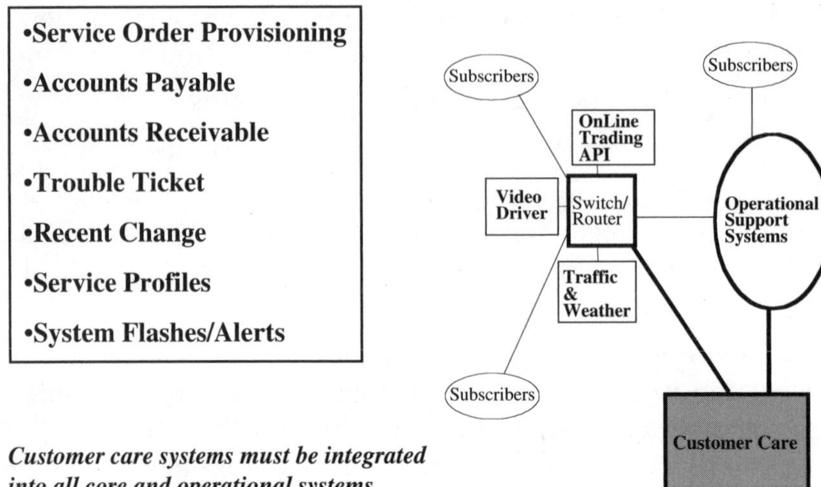

FIGURE 7-5 Customer care systems

INTERNAL VIEW

Different departments can operate the various systems that comprise a customer system. However, these systems must ultimately communicate relevant information to the customer care professional in order to address any questions or concerns articulated by the customer. The internal operation of a service provider will vary from provider to provider. The key point to understand about customer care is that it must communicate relevant information to the customer. The challenge for a service provider is that internal firewalls must be constructed in the respective systems in order to enable the responsible department to act on the request for change or service without overwhelming the customer care professional with information that is not relevant to a customer.

For example, a customer reports a trouble. The customer care professional opens a trouble ticket. The ticket is automatically downloaded to the technicians (hardware or software) for action. The technicians incorporate the information in their tracking system. The tracking system will be populated with the name of the personnel assigned to act on the ticket, the time when the ticket was issued, the time when the ticket was closed, the name of the technical supervisor, and specific actions taken by the technician to resolve the ticket. Half of the information listed is not automatically listed by the customer care professional's system. The customer care professional serves as the face of the company and as a means of controlling information in and out of the company.

As noted previously, customer care systems are not single pieces of equipment, but are tightly coupled with an array of other systems. These other systems are as follows:

- **Service order provisioning systems** Provisioning of new service. This system accepts customer orders for a new service or modifications to the existing service. This system is tied to a technical support desk that activates the new service, whether it is a new screen name for an existing account or a new account.
- **Accounts payable** Customer account status. This system is tied directly to the accounting/billing department.
- **Accounts receivable** Customer account status. This system is tied directly to the accounting/billing department.
- **Trouble ticket reporting systems** Customer reporting of troubles. However, the system also reflects system troubles that are not necessarily reported by customers, but instead are reported by employees. A technical support desk is the recipient of all customer complaints. This desk is then tied to a group or organization that works to resolve the user's complaint.
- **System recent change** Customer care professionals are kept apprised of system changes and updates. In the event a customer calls and wants to report a trouble, the customer care professional may be able to explain that the trouble is only a temporary situation due to system changes. Every department acting on either a trouble or a modification to the software or hardware notes this activity in the customer care system. An informed customer is a happy customer.

- **Customer recent change** This system reflects customer changes to his/her own service profile. This system is tied to the technical support desk and accounting/billing departments. This is a line of data on the customer care professional's display screen reflecting changes regarding the customer.
- **Customer service profile** This system reflects the customer's current service profile. It is tied directly to all departments.
- **System flashes** The system flashes are reporting tools of activity within a provider's core switching/system infrastructure. This system is tied normally to a technical support desk or technical organization. Changes in the system of any kind are reflected in the event that service to the customer is affected. Once again, an informed customer is a happy customer.

Figure 7-6 is a depiction of how these support systems are integrated with one another.

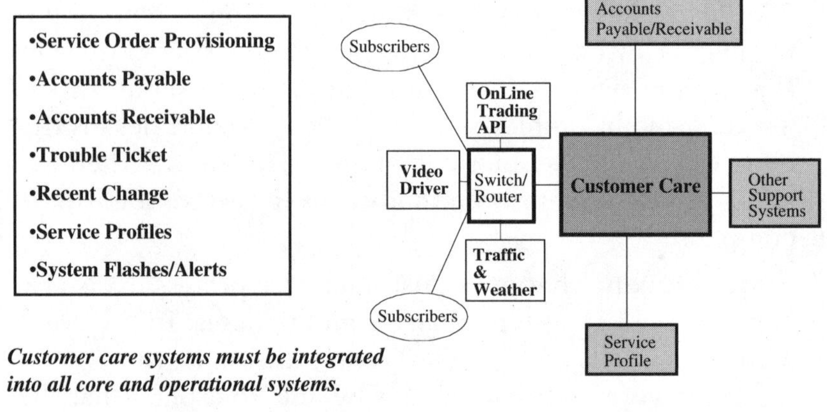

FIGURE 7-6 Support system integration

MOBILITY AND CUSTOMER CARE

Internal communications and operational effectiveness is essential in order for the customer care process to work. The new mobile Internet adds a new dimension to the care and feeding of the customer. Mobile customers obtaining Internet service today are restricted to obtaining it from their wireless provider. However, at some point, the customer will be able to obtain Internet access regardless of whether they are in his/her home system or roaming. Roaming became essential to the growth of wireless. In the mid-1980s, it became clear to the wireless providers that without roaming, the user could not travel out of his/her home system. Without the ability to roam, the usefulness of wireless was dampened. The same will occur in the wireless Internet.

The user will need to have his/her customer needs addressed in a roaming system. It is easy for the wireless carrier to rely heavily on its third party partners to help address questions and concerns. In all likelihood, the wireless carrier will be branding the home Web pages and access/gateways/portals with its name, whereas the actual content site will be provided by some third party that will co-brand the content site with the carrier. The point is that the wireless carrier can reduce its customer care role in the wireless Internet by focusing only on its primary access points/pages. Figure 7-7 illustrates this concept.

CUSTOMER CARE: CHALLENGE

Customer care is undergoing changes throughout the telecommunications industry. Much of customer care is outsourced to third parties claiming to know the provider's business. The cost of customer care is driving ISPs to outsource. The bulk of the cost for maintaining an internal and fully staffed customer care department is in the people. The less one pays, the less talent the ISP gets. In other words, "you get what you pay for."

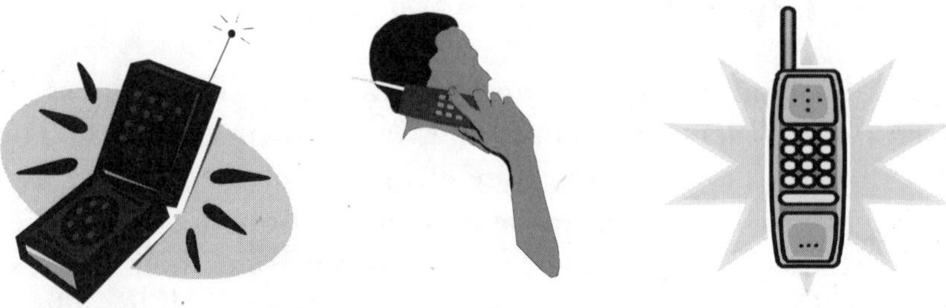

- **Roaming**
- **Branding that crosses borders**
- **Easy Access**
- **Open Access**
- **Standard user interface and procedures**

FIGURE 7-7 Wireless Internet: Mobility and customer care

Outsourcing customer care has its advantages on the operational side. Typically, outsourced firms tend to serve a wide breadth of customers (that is, ISPs) and therefore, have gained a great deal of experience in dealing with a wide variety of customer personalities and questions. This can only help the ISP. The challenge is the automation of customer care. Figure 7-8 is an illustration of customer care automation.

Many ISPs rely on both e-mail to the technical support desk and menu-driven telephone call-in desks. The problem is that as the complexity of communication increases, the more the customer will want to speak with a person. In the Internet service provisioning business environment, early adopters and technical people will find using an automated system great, different, and even fun to use. However, the novelty is unseen with most users and eventually, even the early adopters and technical people will find the novelty of an automated system tired after one of their complaints has gone unresolved in a quick fashion. Most ISPs have found that heavy e-mail users or Web surfers have a tolerance for unresolved complaints within a few

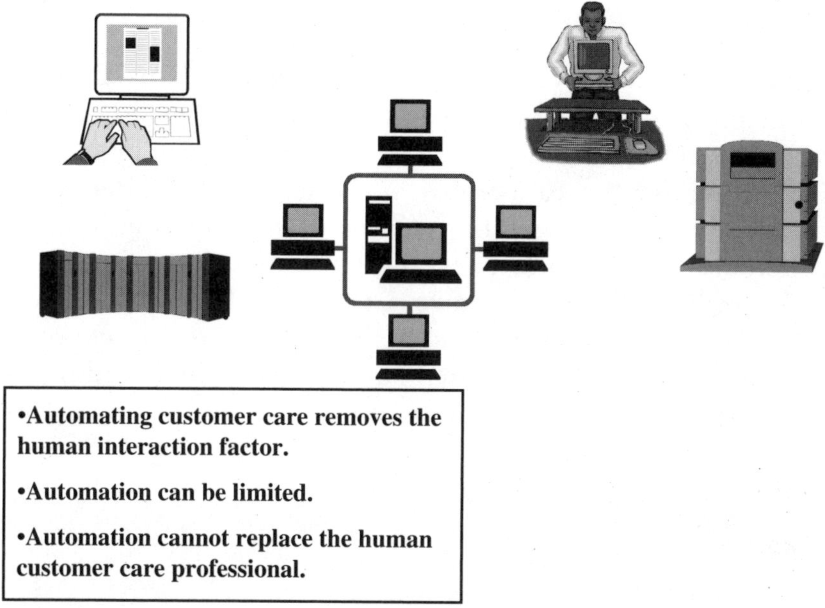

FIGURE 7-8 Customer care automation

hours. The customer care challenge is to maintain a cost effective and useful customer care center without losing sight of the customer. In regard to customer care effectiveness, a direct correlation exists between customer care and subscriber growth and retention.

Customers do not do business with providers that do not treat the customer with courtesy and respect. Often times a poorly trained customer care representative will say all of the wrong things, which angers the customer and results in the loss of business. One should remember to place him/herself in the shoes of the poorly served customer in order to understand the needs of the customer.

The next section will deal with the issue of billing. Billing is the next touch point related to customer care. This is the reason for including customer and billing together in this chapter.

BILLING

Billing is a generic term. However, billing as it is used by the layperson is comprised of two activities: recording and bill rendering. Recording concerns the recording of all call or transaction details related to a user's network usage. Bill rendering involves the processing, rating, bill creation, rendering, and collection of money associated with the network usage.

Recording of network usage data for the ISP would appear relatively simple; however, a network time usage count is not that simple. ISPs must maintain an accurate record of the following:

- The user's real name
- Password
- Logon name
- Multiple logon names (if any)
- Credit card account number
- Date and time stamp—start and end
- Service profile

Mistakes are made far more often than one can imagine. Maintaining a correctly correlated record of usage and user profile is an enormous undertaking. Creating a bill or invoice that makes sense and is correct is also an enormous undertaking. The collected information is used for corporate revenue reporting as well as customer bill creation. The information is also used to support activities related to network usage leased from other carriers. It is also used to support all revenue assurance investigations. More discussion on revenue assurance will continue later. Figure 7-9 is a depiction of how call/activity detail information is used.

The most common mistakes of customer billing are as follows:

- **•Corporate revenue reporting**
- **•Financial analyses**
- **•Future growth projections**
- **•Revenue assurance**
- **•Marketing studies**
- **•Customer behavioral studies**

Billing databases gather and record a myriad of customer information and call related information.

FIGURE 7-9 Call/Activity detail information uses

- Incorrect rates—Unlimited monthly versus fixed time monthly.
- Incorrectly reported paid usage versus free usage—Some ISPs make claims of "zero charges" for entering certain Web sites. The question is, how many people actually check the amount of time they are in the free usage site?
- Incorrectly spelled names.
- Incorrect addresses.
- Incorrectly listed debits and credits.

Figure 7-10 is a rendering of common billing errors.

The previously mentioned errors were so prevalent in the earliest days of ISPs that they ended up moving to signature-on-file accounts and flat rate accounts. In other words, the ISPs charged the customer's credit card directly as opposed to sending him/her a bill. Signature-on-file accounts enable the ISP to charge a credit card without ever having to send a customer a detailed bill in which the ISP would have to explain every charged item or event line item. Signature-on-file customers

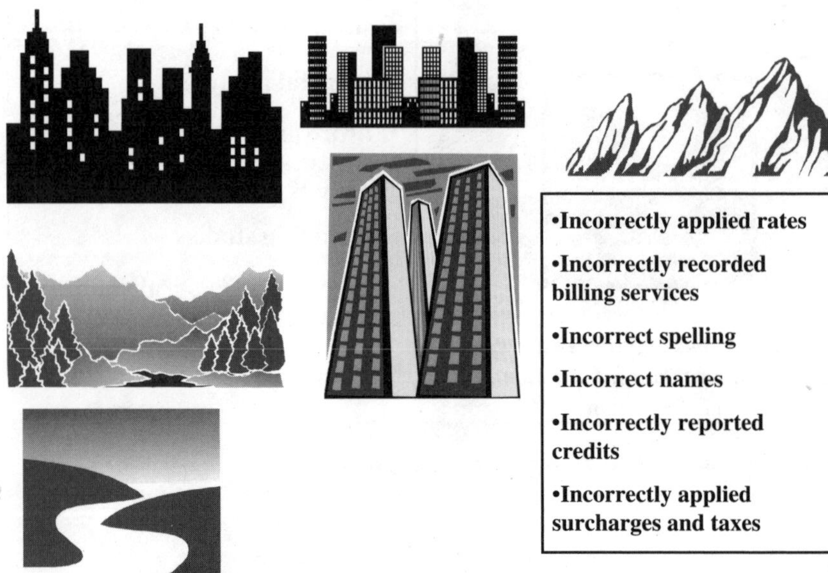

•Incorrectly applied rates
•Incorrectly recorded billing services
•Incorrect spelling
•Incorrect names
•Incorrectly reported credits
•Incorrectly applied surcharges and taxes

FIGURE 7-10 Common billing errors

simply pay the bill and do not ask questions; if the bill looks reasonable, they pay it. Very few people do an item-by-item check of their credit card bills. If it were not for signature-on-file accounts and flat rate billing, *Voice Over IP* (VoIP) would present numerous challenges for the ISP. Users have become accustomed to detailed bills for voice.

Billing Costs

Customer billing is a necessary cost center. The activity does not cause one to want to buy a service; rather, billing, if done properly, brings the revenue into the service provider. The traditional costs of billing are:

- Labor.
- Computer processing—Involves identifying the type of usage and assigning rates to the usage.

- Product development—Rate plan creation and maintenance.
- Paper.
- Bill format/presentation—Large amounts of time and money are spent to create a suitable way of displaying information that accurately communicates the billing information without confusing the customer.
- Cost of mailing.
- System testing.
- Collection process—The check or credit card transaction must be processed.
- Record storage.

Figure 7-11 is an illustration of the costs associated with customer billing.

Costs are associated with all of those activities. Electronic billing over the Web does not necessarily remove all of the aforementioned obstacles. The cost components are the same, but different in degrees of intensity/emphasis. With flat rate plans and signature-on-file, the costs associated with electronic billing are:

- Labor.
- Computer processing.
- Product development—Rate plan creation and maintenance.
- Bill format/presentation.
- System testing.
- Collection process.
- Record storage.

Customer billing is not simple, but is essential.

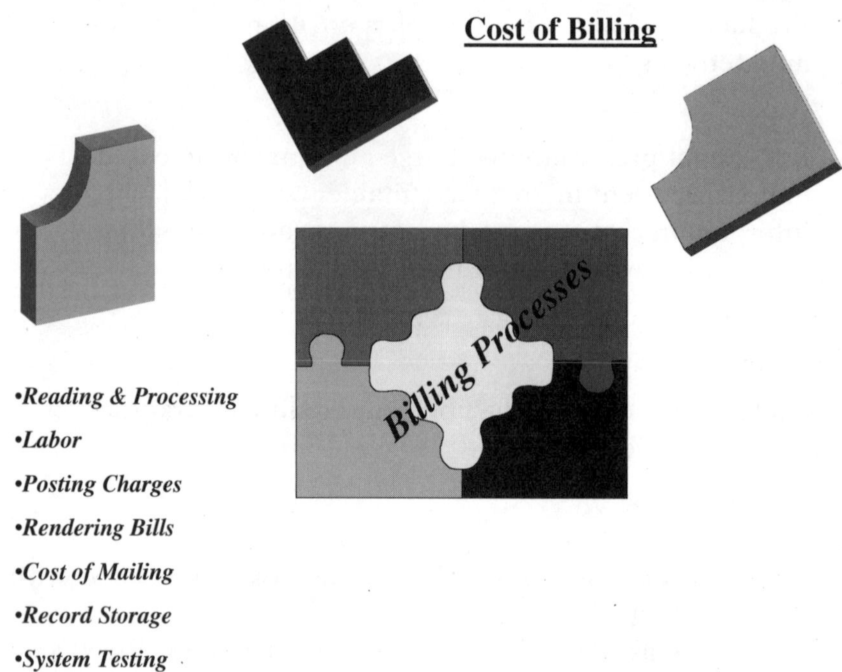

Cost of Billing
Billing Processes
•Reading & Processing
•Labor
•Posting Charges
•Rendering Bills
•Cost of Mailing
•Record Storage
•System Testing

FIGURE 7-11 Cost of billing

OPERATIONAL SUPPORT SYSTEM (OSS) INTEGRATION

Integrating the functions of the various operational support departments is a major task that cannot be trivialized. The lack of "back-office" support has destroyed voice and data CLECs (ISPs). The inability to monitor and manage the network has left customers without service for hours. The inability to properly bill has resulted in revenue leakage for the ISP. The existing wireline and wireless carriers have spent years working to master their own *operational support systems* (OSSs) with many failures. In fact, many, if not all of the startup Internet companies, have neglected to address OSS because it is considered a cost, not a profit generator. This is an improper and dangerous view to hold. Hundreds of cases since 1998 have occurred where ISPs were unaware of their own network failures and unaware of whether their customers were being properly billed. In some cases, people had service even though the ISP billing department noted that due to lack of payment (for a flat rate service), all service had been discontinued. In other cases, service would be disconnected due to non-payment of bills when the accounts had not been properly credited. Part of this was due to lack of the appropriate systems and in some instances, the situation was due to a complete lack of communication between OSSs.

Conceptually, OSSs are comprised of the following systems:

- *Network Management Systems* (NMSs)
- Service Management Systems—service provisioning
- Customer care systems
- Billing systems

Integration of these systems involves the creation of common platforms that enable the easy flow of information between all of these systems. Customer care systems receive data from a variety of support systems involved in the activation and maintenance of a customer account. Much of this information is transmitted between systems via personnel who physically type information into a variety of display screens to which multiple departments have access. A fully integrated OSS enables the automated flow of information from one system to another. In such a case, a user would type information into a screen regarding a specific customer order. The inputted information would automatically be merged from the inputted system to all associated systems. Figure 7-12 depicts the nonintegrated and integrated OSS.

Integrating the OSS would ensure a flow of information throughout the ISP operation. It assumes total intersystem interoperability. Communication between systems is seamless. Figure 7-13 illustrates the concept of integration. The transfer of information at the presentation level appears seamless. However, integrating OSSs is currently an arduous task.

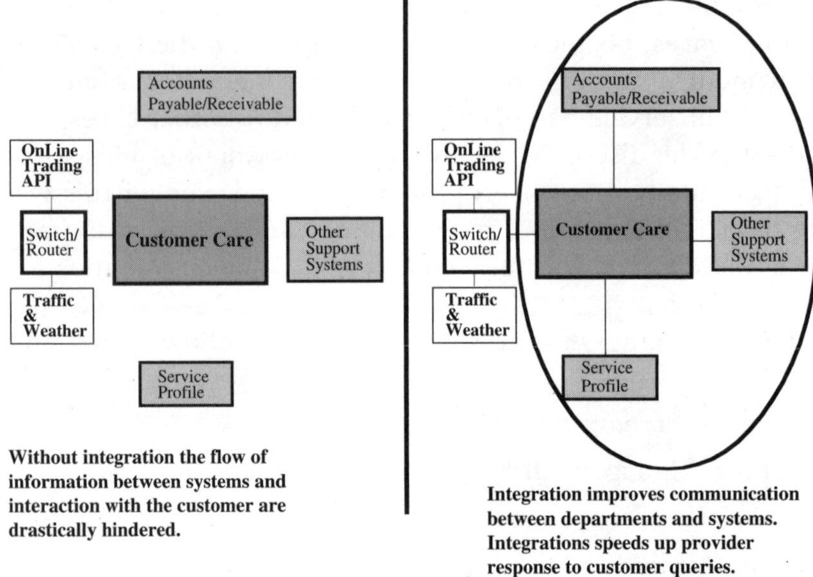

FIGURE 7-12 The integrated and nonintegrated OSS

Integration improves communication between departments and systems. Integrations speeds up provider response to customer queries.

Seamless communication between systems and departments is nearly impossible given the plethora of vendor operational support systems.

In reality, Customer Care integration is slowly being implemented throughout the industry.

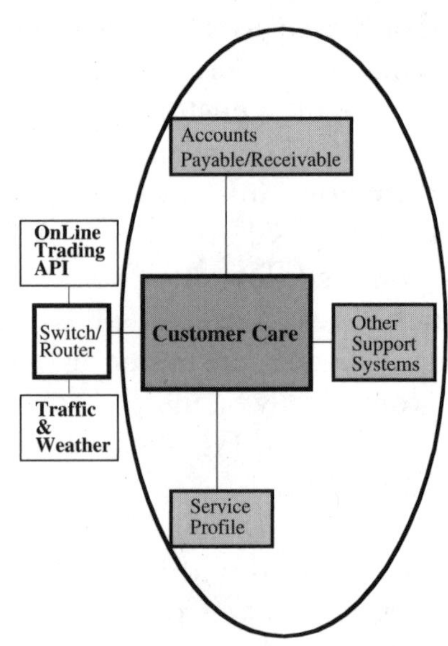

FIGURE 7-13 Interoperability and Seamlessness

Different vendors manufactured most of the OSSs in place today. No thought was given to ensuring intersystem communication.

A variety of mediation devices are in the marketplace today. Mediation involves the provisioning of a network element that communicates to all of the back-office systems. In a sense, the mediation device is an interoperability device or gateway communicating to multiple systems. Mediation is a step toward integration. It supports the following:

- **Intersystem communication** Mediation systems facilitate communication of information between different operational support systems.
- **Single point of control and system access** A mediation system supports a single point of control and access to all interconnected support systems.
- **Facilitates human interface to the systems** A good mediation system is user friendly.
- **Information availability between support systems** A mediation system not only receives information, but it communicates information to the interconnected OSSs.
- **Measures individual system metrics** A mediation device monitors system health and measures system performance.
- **Operational information analysis** Because a mediation system measures and records operating data, it serves as an ideal analysis tool.

Figure 7-14 is an illustration of the mediation concept. Mediation is not integration; it is an interim step toward integration. Integration involves the use of single platforms and single programming languages. Mediation enables existing systems (legacy systems) to continue operating without change.

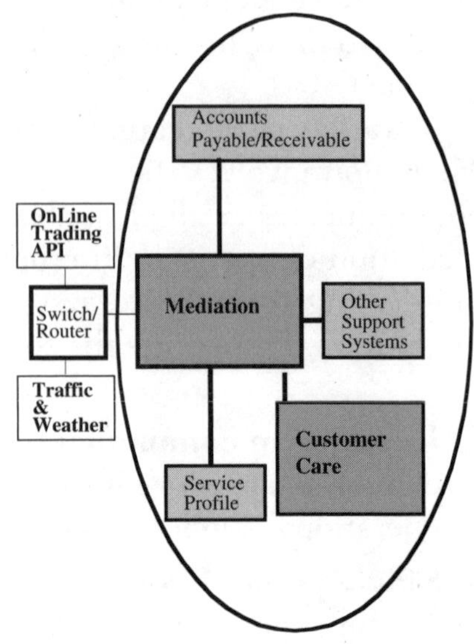
•Mediation serves as an interim step toward a fully integrated OSS approach.
•Mediation devices can meter system performance information.
Accounts Payable/Receivable
OnLine Trading API
Switch/ Router
Traffic & Weather
Mediation
Other Support Systems
Customer Care
Service Profile

FIGURE 7-14 Mediation

REVENUE ASSURANCE

Revenue assurance is a company-wide process in which every process is reviewed for effectiveness. Revenue assurance is an art in which a provider is seeking to maximize revenues via effective management practices that affect the entire revenue process. The revenue process includes the following activities:

- Marketing strategies
- Sales strategies
- Customer care
- Provisioning
- Accounts payable
- Accounts receivable
- Network planning
- Network operations

Revenue assurance is focused on protecting the revenue stream by ensuring that money (revenue) is not spent wastefully on inefficient processes. Revenue assurance is customer centric. Without customers, no revenue exists. Figure 7-15 illustrates the revenue assurance flow.

Each and every one of the aforementioned areas can cause revenue leakage throughout the company. Internal auditors focused on operations efficiency examine the operating procedures of each department and make the appropriate recommendations to improve departmental operations. Revenue leakage is not just about operations efficiency. Revenue assurance is different for different providers. Some providers will focus on different ways of plugging revenue leakage. Figure 7-16 is a rendering of how revenue assurance can affect a company.

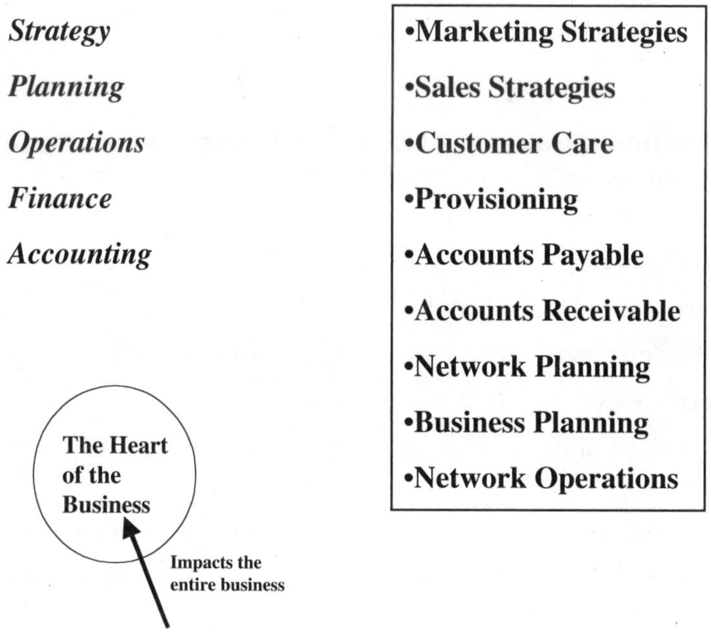

FIGURE 7-15 Revenue assurance

•Improved financials
•Improved revenue collection
•New marketing strategies
•New sales strategies
•Identify operational problems
•Improved customer care
•Improved external communications
•Improved emergency response
•Improved overall efficiency

FIGURE 7-16 The effects of revenue assurance

Revenue assurance is a process that can result in the following:

- Changes in billing systems to ensure data integrity.
- Redefining the billing for a product.
- More cost efficient ways of delivering the service.
- Better ways of validating transactions.
- More efficient and faster credit worthiness checks of the users.
- Revenue reports for a service can result in identifying new potential target markets, which can result in new marketing strategies.

- New marketing strategies can result in re-examining current business relationships with other companies.
- Customer care problems may be identified.
- Service provisioning may be identified as a trouble. Service orders may be processed incorrectly resulting in customers not receiving the correct service. One example is the individual who expects a monthly flat rate rather than hourly usage-based service. Another example could involve incorrectly restricting a screen name. Another example could involve misidentifying access to a free site as a paid site.
- Sales departments can have Web pages redesigned to better communicate the product message.
- Web site access can be improved to enable faster access.

Revenue Assurance and Fraud

Revenue assurance is often associated with fraud prevention. Whole departments have been established to combat customer fraud. Fraud can take on many forms. It is a broad term to define illegally obtained service. Fraud can encompass the following:

- False credit card numbers.
- False customer identities.
- Rotating customer accounts—Fraudulent users may rotate through a variety of stolen credit card numbers in order to hack into Web sites for the sole purpose of stealing products online.
- Hacking.
- Identity theft.
- Simply taking advantage of every operational hole a customer can find in the ISP's network.

Fraud is not a trivial matter. In the early days of cellular, it had been estimated that hundreds of millions in dollars of revenue were lost each year to fraud. The fraud had been so severe that many cellular carriers could not even estimate how much revenue was lost. Even the wireline voice telephone network still encounters fraudulent users. The fraud situation in the world of the wireline Internet is also quite severe. The wireline Internet has encountered all of these situations just mentioned. One cannot deny the fraud in the Internet industry. In many areas of the country and many other countries, it is a crime to hack.

The Internet crime has become synonymous with "white collar crime." Fraud is a "white collar crime." The use of signature-on-file accounts has made it easier for individuals to commit Internet fraud. The credit card number can be checked quickly; however, an Internet criminal only needs a couple of hours to commit millions of dollars of theft before a stolen credit card is reported.

Revenue assurance is a thought process that should be integrated throughout the provider's business processes. Figure 7-17 stresses the role of revenue assurance.

SUMMARY

The Internet industry rapidly understands the need to focus on the customer. Some Internet companies have recently discovered what bad publicity can do when the public feels it has been misled or lied to. The customer is a king or queen. The customer pays the bills and ultimately pays the salaries of the provider. Processes are needed to not only check the way in which a company does business, but also the way in which a company does not do business. A provider cannot outright accuse a customer of misdeeds. However, an unofficial and unspoken understanding exists between provider and customer

•Improved financials

•Improved revenue collection

•New marketing strategies

•New sales strategies

•Identify operational problems

•Improved customer care

•Improved external communications

•Improved emergency response

•Improved overall efficiency

Revenue assurance is a thought process.

Revenue assurance affects every aspect of a company's operations.

Revenue assurance is both internally and externally focused.

Many service providers will view differently from one another.

FIGURE 7-17 The role of revenue assurance

that gives the provider the right to check a customer's credit worthiness. Customer care, billing, and revenue assurance are multifaceted internal and external processes that are focused on the provider-customer relationship.

Whether Internet access is wireline or wireless, the basics of doing business with a customer are the same. The unglamorous world of customer care and billing are essential components of the glamorous world of the Internet. At the end of the day, the Internet is like any other business, and the mechanisms used to manage the business are the same.

The next chapter will summarize the positions taken in the book and take a closer look at the evolution of the Internet.

CHAPTER 8

THE INTERNET BUSINESS MODEL: THE FUTURE AND ITS ECONOMICS

The Internet is not a product or a service. The Internet is a different thing to different people. It is a medium for change in the telecommunications industry. The Internet is transforming the telecommunications industry into an open information medium.

CONVERGENCE AND THE OPEN PIPE

The Internet is the embodiment of convergence. It is an open pipe of information that enables access to and from anyone who uses a computer or other terminal device. No single technological breakthrough in the last 20 years has caused such excitement in the computing and telecommunications industries. However, the Internet is under attack because of those individuals who misuse the medium for nefarious schemes and acts resulting in personal, emotional, and physical assaults. The Internet did not cause the problems per se, but it provided the avenues to fuel the resultant actions. One can find out how to make a bomb, design a gun, spread hate messages, torture, and even kill. The Internet was meant to enhance, not hurt communications. Due to these abuses, governments around the world are seeking ways of controlling the Internet and its contents.

Government control of this resource as advocated by numerous lobbying groups can be destructive to the growth of the Internet. A balance must be maintained that allows for the creativity of the business and technical communities, which cause the Internet to grow in capability and market penetration. Too much control can be stifling, but not enough control can be just as detrimental.

The Internet is an issue for society. As cliché as it may sound, this man-made creation has permeated all aspects of society and all nations within a few short years. This book is not about morality. It does not purport to know all or proclaim to even know a great deal about what is morally correct for society. However, it is understood that the Internet has been used to hurt and illegally profit. Groups across the globe are working to find ways of securing the safety of children via restricted access methods. Groups that spread hate messages are being forced into bankruptcy by court actions in the hope of stopping the spread of hate messages. Consumer advocacy groups now surf the Web for scams that bilk the elderly and others out of a lifetime of savings. Individuals set up their Web sites to spread the word about something bad or something good. These individuals are creating public dialogue on a variety of issues.

The Internet possesses both positive and negative attributes. A typical engineer's point of view would be, "The people are the ones to blame, not the technology."

Internet professionals need to stand up and take charge of how the Internet is used. If the various nefarious and criminal users of the Internet are allowed to continue unmonitored, the government will step in and heavily regulate the medium. As it stands now, a great deal of activity is present to regulate the Internet via government legislation. However, many of the government measures can be easily bypassed. The "open" attribute of the Internet makes it difficult to stop people from launching negative Web sites. The only way to control the content of the Internet is to prohibit individuals from participating in the delivery of Internet content. In short, make it illegal for non-government certified providers to provide any kind of content or Internet access. This step would be an extreme measure that can be avoided if the industry takes control of the matter at hand with self-regulation.

The industry is at the cusp of explosive growth. If the industry does not take matters into its own hands, the commercial growth that experts expect will not occur. The Internet grew, partly, because it was not regulated. If the government heavily regulates the operation of the Internet, then we may see investors abandoning the market faster than they are doing so already. The result could be limited industry growth.

Any Internet provider providing voice service should be held accountable to the same standards that the existing wireline carriers are held to. Figure 8-1 is an illustration of the Internet and its impact on society.

THE PUBLIC RIGHT

In the wireline telephony world, voice service is perceived as a public right. The state and federal regulatory agencies, and even the customer hold this view. This dates back to the wireline telephony days in which the telephone was the principal means for all people to communicate over distances. Government regulations in Canada and the United States regulate a wireline telephony environment in which the local telephone company is held accountable and in some instances, criminally liable for communication failures that result in injury or death.

In the early 1970s, a large United States city on the East Coast suffered communication problems so severe that many people could not get a dial tone. One year, a fire occurred that resulted in loss of service for tens of thousands of subscribers. This resulted in several hospitalizations, government investigations, threats of criminal prosecution, and eventually the dismissal of several telephone company executives. In the end, the industry discovered to their happiness and dismay that the telephone was now perceived to be a public right. This 1970-ish

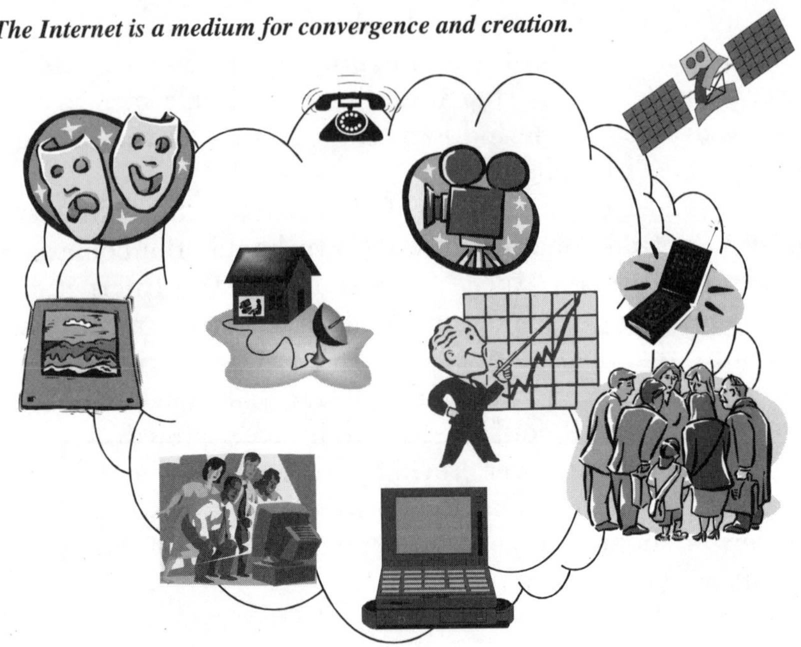

The Internet is a medium for convergence and creation.

FIGURE 8-1 The Internet—Society and convergence

service disaster resulted in a public outcry that reshaped the thinking of the entire telephony community. Forty years ago, one could find people who could remember when the world did not have the telephone. No one alive today can say that they lived in a world where a telephone did not exist. The Internet will become a common way of communicating like the telephone. The Internet notwithstanding will be held to the same unofficial standard (customer perception) or expectation if it becomes the primary medium for voice. The motto for voice service provisioning companies ought to be "service, service, service."

The voice is the most basic form of telecommunications service. If an ISP decides to provide voice as one of its anchor products, it must be prepared to deal with the public perception set by the wireline telephony world. E-commerce and m-commerce will not save any ISP from bankruptcy if the voice service is mishandled so badly that the user believes the provider is incompetent in the provisioning of voice.

ISPs wishing to provide voice as part of their service offerings will need to meet federal and state regulatory commitments. Once voice becomes part of the ISPs' service offering, the government agencies will demand that the ISP be classified a CLEC. Some regulation will be required if the public expects quality and protected treatment. Each state regulatory agency should be able to provide safety for children surfing the Web. Today, state regulatory agencies ensure fairly priced, quality service for the subscriber. The wireline and wireless carriers must support a number of federal commitments. The Internet is a medium that deserves some kind of regulation, but definitely not the same kind of federal and state regulation to which the wireline telephone companies are subject.

However, an aspect of the Internet exists that the voice-only world does not have to deal with. This aspect is already being dealt with by the legislation: protecting children. Unlike the telephone or wireless handset, the Internet has become a way for the criminal to lure and emotionally abuse children. In some cases, the evilness has gone as far as pornography involving children. Regulation has already begun with laws being passed to protect children.

Given all of the negative publicity the Internet has received, as evidenced by a number of horrible and unfortunate incidents resulting in the shooting deaths of children and adults, one should expect some form of regulation of the industry. Some of the steps taken by the industry to self-regulate have resulted in the following:

- Restricted access by children, which can be done by assigning restrictions based on screen name.
- Self-monitoring ISPs—ISPs that monitor the types of sites being accessed by users. This is done at the request of parents.
- ISPs refusing to display Web pages of undesirable Web sites.

In the United States, a carrier must comply with other commitments in order to be a wireline or wireless carrier. These commitments are laws that were passed outside of the Telecommunications Act of 1996, but still carry the same/equal weight of legal obligation as the Act of 1996. These acts include *Communications Assistance for Law Enforcement Act* (CALEA) signed into law in 1994, FCC rules for Ensuring Wireless Carrier Compatibility with Enhanced 9-1-1 Emergency Calling Systems (FCC Docket No. 94-102), and the Universal Service Fund. The CALEA law and the FCC rules on wireless carrier location for 9-1-1 are requirements for all carriers entering the marketplace. The Universal Service Fund as redefined by the Act of 1996 has the potential of having the most far-reaching impact at every income level in the Untied States.

COMMUNICATIONS ASSISTANCE FOR LAW ENFORCEMENT ACT (CALEA)

In October 1994, Congress passed and the President signed the CALEA. The law was designed to position law enforcement to better respond to the rapid advances in telecommunications technology. The law was also designed to eliminate obstacles faced by law enforcement agencies in conducting electronic surveillance. Electronic surveillance is defined as "both the interception of communications content (wiretapping) and the acquisition of call-identifying information (dialed-number information) through the use of pen register devices, traps, and traces." The rapid advancement of communications technology, the growth of new types of services, and the increase in the number of carriers has complicated the work of law enforcement to the extent that surveillance is hampered.

Since 1970, U.S. telecommunications carriers have been required to cooperate with law enforcement personnel in conducting electronic surveillance. CALEA took this one step further requiring telecommunications carriers to modify and design their switching systems and other support equipment, facilities, and services to ensure that authorized electronic surveillance can be performed.

The CALEA law affects wireline, cellular, and PCS carriers. The question now is, how will the Internet be affected, especially in the mobile market?

CALEA requires telecommunications carriers to ensure that their equipment, facilities, and services will meet four functional, or assistance capability, requirements that enable law enforcement to conduct authorized electronic surveillance. These requirements are as follows:

- Carriers must be capable of expeditiously isolating and enabling the government to intercept all wire and electronic communications within that carrier's network to or from a specific subscriber of such a carrier.
- Carriers must be capable of rapidly isolating and enabling the government to access call-identifying information that is reasonably available to the carrier, except to the extent that the telephone number alone may determine the location.
- Carriers must be capable of delivering intercepted communications and call-identifying information to a location specified by the government, other than the premises of the carrier.
- Carriers must be capable of conducting interceptions and providing access to call-identifying information unobtrusively. Furthermore, telecommunications carriers are required to protect the privacy and security of communications and call-identifying information not authorized to be intercepted, as well as information concerning the government's interception of the content of communications and access to call-identifying information.

WIRELESS 9-1-1

The FCC rules for wireless telecommunications calling party location in support of 9-1-1 emergency calls were passed in 1994 and reaffirmed in 1996 and 1999. The FCC rules on wireless 9-1-1 require wireless carriers to provide the location of the party calling for help in a two-stage process. Stage 1, known as Phase 1, requires the wireless carrier to provide the calling party's telephone number, cell site, and sector location (the cell site and sector from which the emergency call is being made). Phase 2 requires the wireless carrier to provide the latitude and longitude of the calling party upon initiation of a 9-1-1 call. Issues regarding liability and cost recovery (for the installation and operation of these systems) have been continuing for a number of years. The deadline to comply with the FCC's rules is October 1, 2001.

This mandate is not going to change whether or not the wireless carrier is providing Internet access. The wireless carriers understand their responsibility in this area.

UNIVERSAL SERVICE FUND (USF)

The *Universal Service Fund* (USF) was established under the direction of the FCC in support of the Telecommunications Act of 1996. The Act redefined the universal service policy as it had been defined by the Communications Act of 1934, the forebear of the Telecommunications Act of 1996. The Act of 1934 had created the FCC and noted the following:

The purpose of the Act of 1934 was " . . . to make available, so far as possible, to all the people of the United States, without discrimination on the basis of race, color, religion, national origin, or sex, a rapid, efficient, Nation-wide, and world-wide wire and radio communication service with adequate facilities at reasonable charges, for the purpose of the national defense, for the purpose of promoting safety of life and property through the use of wire and radio communication, and for the purpose of securing a more effective execution of this policy by centralizing authority heretofore granted by law to several agencies and by granting additional authority with respect to interstate and foreign commerce in wire and radio communication, there is hereby created a commission to be known as the 'Federal Communications Commission,' . . . ."

The Act of 1934 defined universal service as consisting of the following principles, as quoted in the Act of 1934:

> " . . . and the Commission shall base policies for the preservation and advancement of universal service on the following principles:
>
> (1) QUALITY AND RATES Quality services should be available at just, reasonable, and affordable rates.
>
> (2) ACCESS TO ADVANCED SERVICES Access to advanced telecommunications and information services should be provided in all regions of the Nation.
>
> (3) ACCESS IN RURAL AND HIGH COST AREAS Consumers in all regions of the Nation, including low-income consumers and those in rural, insular, and high cost areas, should have access to telecommunications and information services, including interexchange services and advanced telecommunications and information services, that are reasonably comparable to those services provided in urban areas and that are available at rates that are reasonably comparable to rates charged for similar services in urban areas.
>
> (4) EQUITABLE AND NONDISCRIMINATORY CONTRIBUTIONS All providers of telecommunications services should make an equitable and nondiscriminatory contribution to the preservation and advancement of universal service.
>
> (5) SPECIFIC AND PREDICTABLE SUPPORT MECHANISMS There should be specific, predictable and sufficient Federal and State mechanisms to preserve and advance universal service.
>
> (6) ACCESS TO ADVANCED TELECOMMUNICATIONS SERVICES FOR SCHOOLS, HEALTH CARE, AND LIBRARIES Elementary and secondary schools and classrooms,

health care providers, and libraries should have access to advanced telecommunications services as described in subsection (h).

(7) ADDITIONAL PRINCIPLES Such other principles as the Joint Board and the Commission determine are necessary and appropriate for the protection of the public interest, convenience, and necessity and are consistent with this Act."

The Telecommunications Act of 1996 redefined the concept of universal service and expanded it to include access to advance telecommunications services and all kinds of information services. All regions and income levels of the United States are required to be included. Regardless of the economic level of the region, every person, child, school, health care facility, and library shall be provided the same level and type of telecommunication and information service provided in areas that have the economic infrastructure to support the most advanced information services.

In order to support this government initiative, a fund was established by the FCC and administered by the *National Exchange Carrier Association* (NECA) to pay for the installation of computers in classrooms and subsidize carriers in low-income rural or urban areas to enhance the telecommunications infrastructure. This fund is paid by all wireline LECs. Some individuals call it a "government entitlement;" however, this is not an entitlement program. This fund is a necessity to ensure the education of the nation's young and the economic viability of the nation itself. Payments into the fund are based on a fixed formula that requires every carrier to pay a portion of its gross revenue into the fund.

The Internet has the potential of becoming bigger than the existing wireline and wireless voice-only network services. The Internet carrier may have to step up "to the bar" and acknowledge some level of responsibility to serving the public well by paying into the USF. However, the mechanism that one would use to measure the Internet carriers does not exist.

THE PUBLIC AND REGULATION

This section of the chapter is not meant to cause a movement toward regulation. Rather, it is meant to educate those Internet players entering the market. The minute the ISP enters the marketplace, he/she has an implied responsibility to the public. This is how existing non-Internet carriers run their businesses. These carriers understand that they possess some level of obligation to the public and whether right or wrong, the customer perceives it as his/her right. For those people who have lived through service disasters, it is clearly understood that this perception dominates all thought and action the minute a taxpayer says the loss of service has resulted in a loss of business or resulted in injury. The next thing the service provider hears is requests for public explanations in front of a variety of government committees.

THE INTERNET BUSINESS MODEL

As noted in Chapter 2, "Business Models," the Internet can be classified into two broad categories: access and ISP. The categories are meant to help filter out the noise of Wall Street's hype about the uniqueness of the various Internet plays.

The *Internet Service Provider* (ISP) is the company that provides the service directly to the user/subscriber. The ISP can be further broken down into sub-categories of ISP types. ISPs provide e-mail, Web surfing, limited file storage, and chat rooms.

Access is a capability that lies at the heart of the Internet business. This access capability is what makes the Internet a valuable tool. Information can be obtained on almost any topic. Much of the database access that is being touted in the trade journals has been focused on commerce, specifically e-commerce and mobile e-commerce. Doing business across the Internet has proven to be a primary driver of the Internet's growth.

Figure 8-2 is a rendering of the two basic categories of Internet companies.

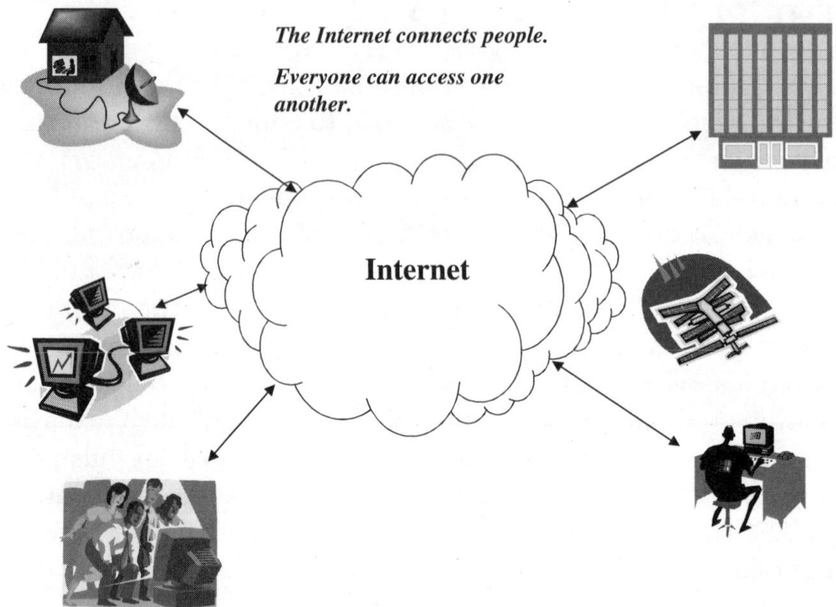

FIGURE 8-2 The ISP and access provider

E-commerce and mobile e-commerce are activities/applications that make use of the Internet's natural and inherent distribution network, that is, access. Wired and wireless e-commerce has given birth to a number of different business relationships over the Internet:

- *Business-to-Business* (B2B)
- *Business-to-Consumer* (B2C)
- *Business-to-Government* (B2G)
- *Exchange-to-Exchange* (X2X)
- *Taxpayer-to-Government* (T2G)

Figure 8-3 is a rendering of the business relationships.

E-commerce and mobile e-commerce have changed the way people shop.

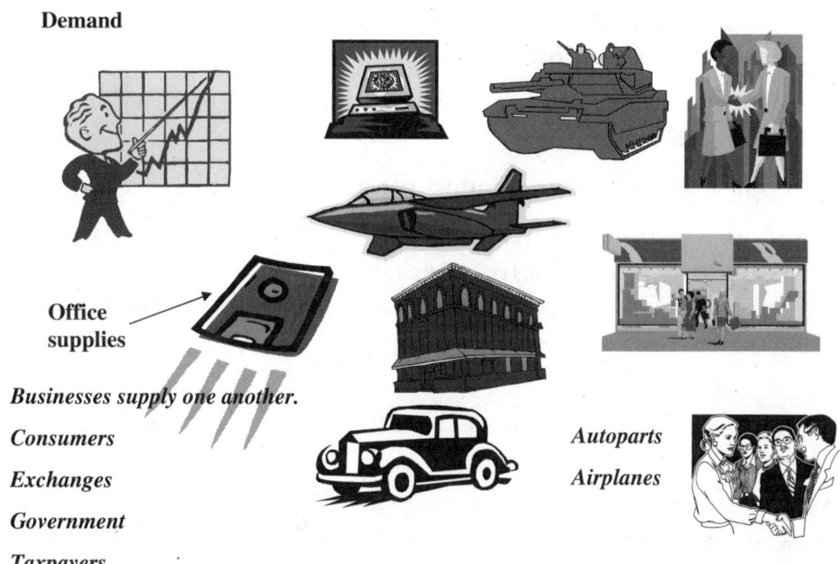

FIGURE 8-3 Business relationships

INTERNET RETAILERS VERSUS BRICK AND MORTAR RETAILERS

Despite the rash of Internet shopping that takes place, a brick and mortar store has distinct advantages over an Internet Web store:

- Physical locations can be passed on the way to work and on the way home from work. One can stop quickly and shop.
- Established brick and mortar retail chains spend billions of dollars on inventory. Internet sites do not. Suppliers tend to listen to the people who spend the most money. Therefore, when supplies are low, the supplier will supply the big spenders first.
- Existing brick and mortar retailers have huge warehouses to store the billions of dollars of inventory it just purchased. The Internet sites have to build these warehouses, which means higher operating costs for storing inventory.

- Brick and mortar stores have a brand presence, whereas Internet sites do not.
- Brick and mortar stores have a human presence, whereas Internet sites do not.
- People like to feel and touch the product.
- Very few people like to wait days for a product to arrive by mail. However, if the Internet site can ease the burden of the act of looking through half a dozen stores to find the right product, why not shop the Internet?

None of this information should give the reader the impression that the brick and mortar store has won the war over the consumer. The Internet shopping sites have advantages over the brick and mortar stores:

- Labor costs are less because salespeople do not have to be paid.
- Lower training costs involved for salespeople.
- No building maintenance costs.
- Advertising costs are lower than the brick stores.
- Product returns are easier.
- One can find lower prices on the Internet for the same product.
- Now it has gone wireless (mobile).

Figure 8-4 illustrates the relationship between the Internet retailer and the brick and mortar store.

The Internet has created an opportunity for the consumer and the telecommunications professional. As of mid-2000, four communities of Internet vendors are available: the ASP, the ERP provider, the CRM provider, and the content provider. These vendor groups are part suppliers to ISPs and part providers to the end user.

Advantages Over the Brick -n- Mortar Store

•Lower labor costs

•Minimal training costs

•Lower advertising costs

•Product returns are easier (in most cases)

The Internet has broadened the marketplace for the retailer.

The Internet has made buying easier: You can do it from home.

FIGURE 8-4 The Web versus the brick

Application Service Provider (ASP)

The Internet industry has even given birth to a new type of vendor: the *application service provider* (ASP). The ASP is a company that writes software that is used on a carrier's network. ASPs have always existed in the telecommunications industry, but they had been hardware vendors until the birth of the Internet.

The ASP in the Internet world writes not only the enabling software, but also provides in some manner content or some type of content-related service. ASPs today are serving as gateways to customer bases. Most of the ASPs claim to have direct access to huge pools of customers either directly or through their own service partners. The services are provided using ASP specific software tools. Another way of describing the ASPs is that they host delivery of content to carriers. The following list includes examples of services that the ADP may provide:

- Internet directory access
- Video and audio content
- Internet radio

- Polling
- Games
- Stock exchange monitoring
- Online purchasing
- Online stock trading
- Weather
- Traffic reports
- Local advertising

The ASP is a value-added vendor for the carriers, such as the Internet carrier or the voice carrier. In a strong sense, the ASP is also an access provider. The ASP provides access to customers for the carrier. One issue exists with the ASP; it is a single application-approach provider. Figure 8-5 is an illustration of the ASP and its role in the Internet.

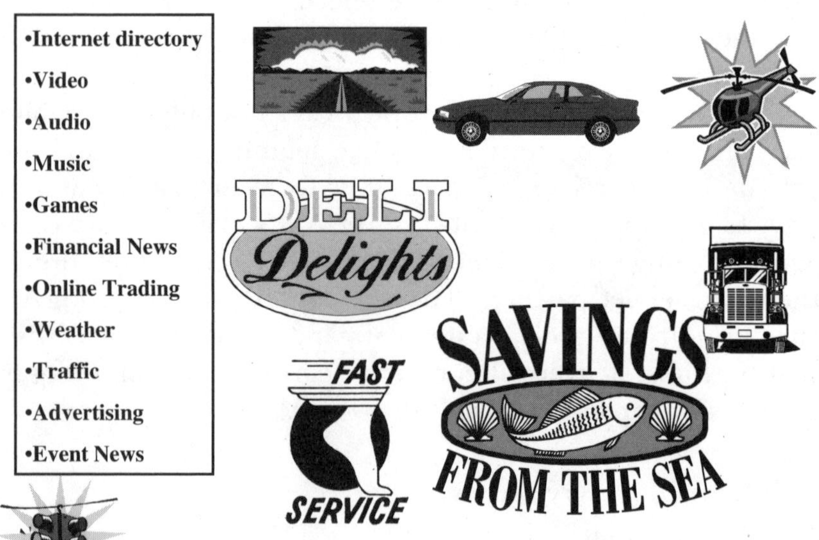

FIGURE 8-5 The application service provider (ASP)

THE ENTERPRISE RESOURCE PLANNING (ERP) PROVIDER

A new type of Internet player is helping to shape the information landscape: the *Enterprise Resource Planning* (ERP) provider. The ERP provider is the ASP plus. The ASP provides a solution to a specific service or set of services, whereas the ERP is a platform that provides operations support, services, and service specific access gateways (portals) for the ISP (wired and wireless). ERP platforms are integrated information systems that can serve all departments or multiple departments within an enterprise (business).

The ERP is a new business segment within the Internet community, but they are a little bit of everything that manage the resources of the client company or ISP. Some ERPs provide software solutions, services, billing services, data warehousing, operations management, Web site hosting, and portal management. Unlike the ASPs, the ERPs have a relationship that is theoretically closer to the wireline and wireless carriers. The ERPs are much like the service bureau and wholesaler in the non-Internet telecommunications environment. However, in the case of ERP, software and hardware integration are components of the business.

The ERP concept arose out of the manufacturing industry where software platforms were used to integrate activities such as accounts receivable, accounts payable, manufacturing processes, order entry, general ledger, purchasing, logistics, and inventory control.

Integration is what sets ERP apart from the old-style service bureau model. An ERP provider is a network within a network. The components of the ERP provider can serve as the main switching and processing elements, and support mechanisms within the carrier's infrastructure. Inside a carrier's business, ERP would support finance, human resources, project management, customer service management, and workflow management. ERP would also support gateway management for the Internet carrier. An Internet gateway functions like a tandem (in the voice wireline world) or a doorway to other networks; it is essentially a traffic control point to different networks. An Internet gateway would manage the information flow/relationship to radically different types of networks. Gateways serve as secure traffic points.

Figure 8-6 is a rendering of an ERP.

The ERP is another twist to the Internet and if one took a closer look at the business relationship, he/she would see that the ERP relationship has existed for many years in the non-Internet world. The ERP and the ASP prove that existing knowledge and experience can be transferred to the Internet medium. Figure 8-7 is another rendering of the ERP model in the Internet business.

CUSTOMER-RELATIONSHIP MANAGEMENT (CRM)

As of this writing, this is a new addition to the Internet business forms. *Customer-Relationship Management* (CRM) is outsourced customer interfacing/customer care. This is a specific area of business opportunities for experienced customer care

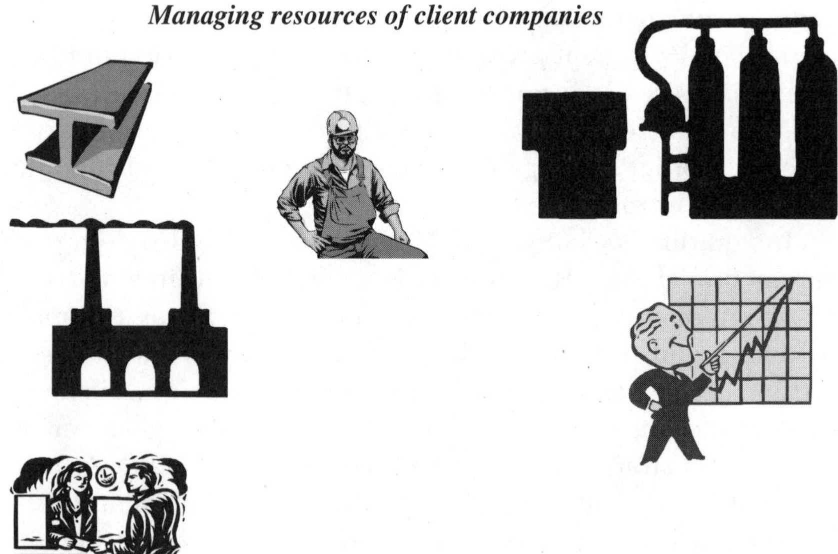

FIGURE 8-6 The enterprise resource provider (ERP)

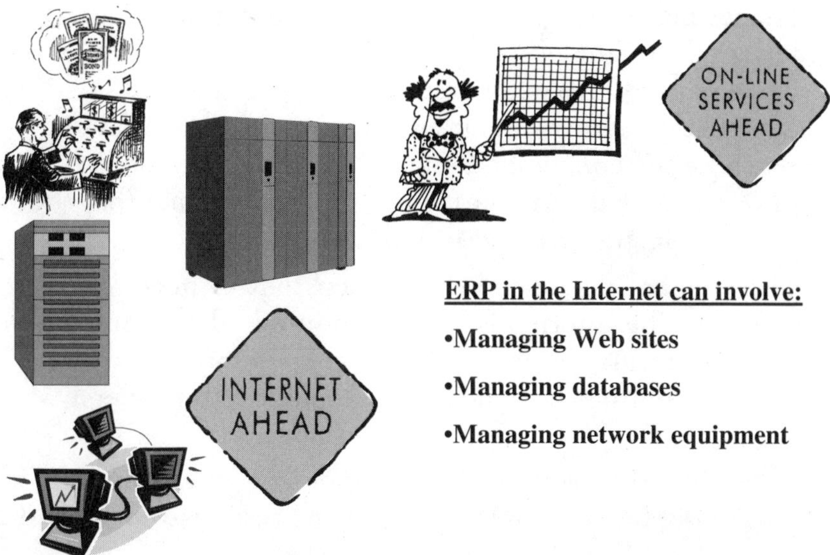

FIGURE 8-7 The ERP in the Internet

people who are looking to transfer their body of knowledge into the Internet arena. The complexity of the customer-to-business relationship has increased due to the Internet. The relationship between the customer and the provider has become more complex due to the following factors:

- The Internet has raised expectations for customers who shop over the Internet. Internet shopping is now synonymous with a low-cost product; get what you want and get it right away. In other words, those who shop over the Internet believe that they can find product quickly, product of superior quality, and product at substantially lower prices.
- Compressed marketing windows—The customer's attention span has decreased due to the immediate nature of the Internet. So many companies use the Internet to advertise that Web surfing has extended into the area of shopping as well.

- Higher marketing costs—The Web does not necessarily lower advertising costs. The need to get the attention of the customer requires companies to spend money on advertising agencies in order to figure out better ways of reaching the consumer. Some of the Internet shopping sites have responded with marketing "pitches" touting that they are more responsive and friendlier.
- The Web has responded with the customers' need for product. The Internet is filled with hundreds of companies that sell a wide variety of products. Therefore, the environment has become more competitive for sellers.

The CRM is a software-based approach to customer care. The thinking behind CRM is optimizing responsiveness to the customer.

As the reader will recall, customer care is simply what the title implies; it refers to every aspect of caring for the customer's questions and needs. Customer care is a process that links the customer directly to the core functions of the service provider. In the case of the Internet, software platforms have been developed to fully automate customer care. In the case of a CRM, the customer fills out his/her own trouble ticket. This particular aspect of CRM does not seem capable of lasting this long. This is taking the concept of the Extranet too far. Customers want to be catered to, not given more work to do. Creating one's own trouble ticket has its benefit. For instance, the MIS professional may find this a quick way of filing a trouble ticket rather than waiting for several minutes to speak with someone. The amount of time spent resolving a trouble over the phone could have been handled more efficiently by sending in the trouble ticket oneself. Extranets are a way for companies to provide a window into their operations and customer support desks from the outside without penetrating any critical areas of the company's operations. Firewalls serve as protective barriers between the company and the consumer.

The problem with a fully automated system is that consumers who are frustrated due to a problem with service want to scream at a person. The upside of a fully automated system is that one does not have to pay or train people to address customer questions and concerns.

The CRM is analogous to the voice driven menu used by telephone companies for customer complaints. Except in the case of the CRM, the complaint is filed over the Internet into the company's Extranet.

Figure 8-8 is a rendering of the CRM.

THE NETWORK SERVICE PROVIDER

The *Network Service Provider* (NSP)is a new term to describe an old concept/existing type of telecommunications player. A NSP provides backbone services to an ISP. The NSP provides an ISP with network connectivity either between the ISP's points of presence or to other ISPs.

Many concepts used within the ISP or Internet realm are repackaged, non-Internet telecommunications ideas. Much of the knowledge and the experience within the non-Internet worlds is

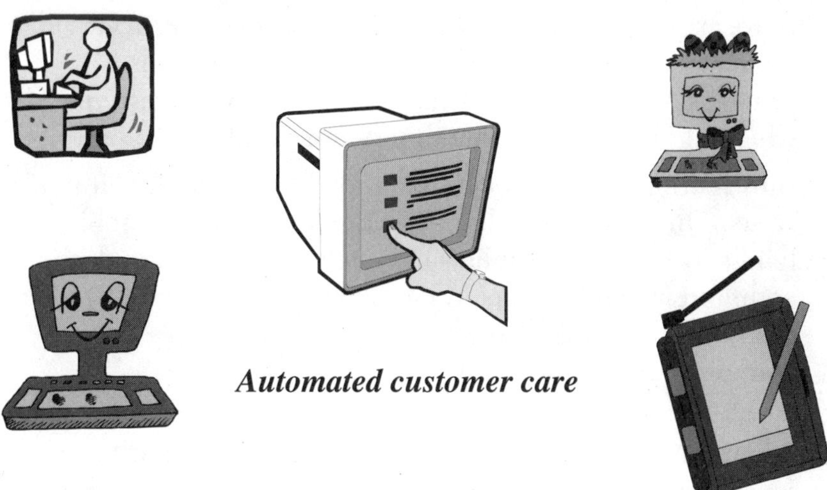

FIGURE 8-8 The customer-relationship management (CRM)

transferable to the Internet world. Figure 8-9 illustrates the NSP concept.

THE CONTENT PROVIDER

To some extent, the ASP and the ERP provide some type of content to the users they serve. Content providers are also sources of material for ISPs and ASPs. Content can be defined as:

- Video.
- Voice.
- Text.
- Telephone numbers.
- Domain names.
- Any information about the call or data file.
- Information that has been manipulated to provide additional value. Data warehousing is one example of information that is mined out of other information. Another example is supermarkets now providing discount cards for either frequent purchases or for sale items. The value of the discount cards is the information filed by the customer to receive the card. This information not only provides buying pattern information about the user, but also provides a wealth of information about buying patterns when all of the cards are examined in total.
- Information that has been enhanced to provide additional value. One example is knowing the location of a mobile handset. First, the public safety value exists. Second, the commercial value exists in knowing where the mobile user is so that directed advertising or even wireless sales coupons can be delivered.

Content is the substance of the service. Content has value because it is about the customer or it is information the customer wants. It only has value if the customer wants to spend

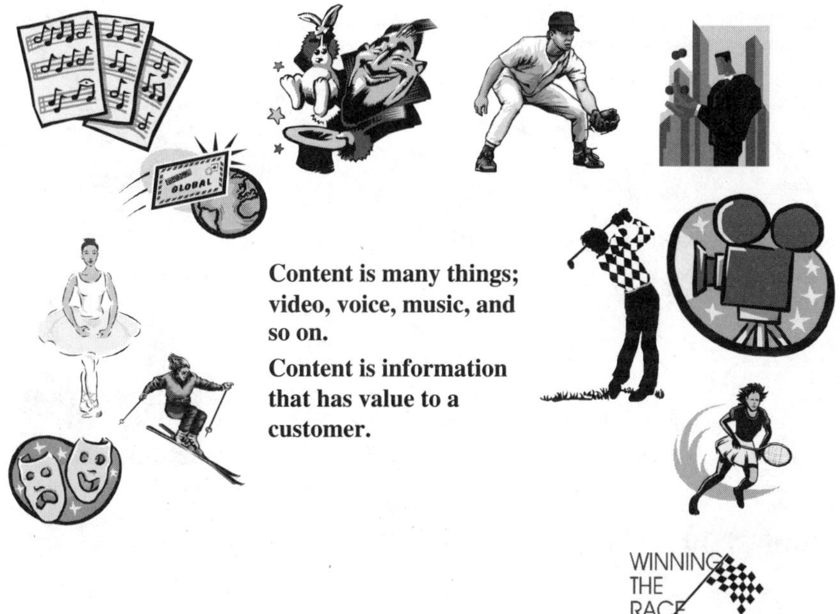

GLOBAL
Content is many things; video, voice, music, and so on.
Content is information that has value to a customer.
WINNING THE RACE

FIGURE 8-9 The network service provider (NSP)

money on it. Figure 8-10 is an illustration of the content provider.

THE OTHER PROVIDER (OP)

A generic statement can be made about the telecommunications industry when it comes to the types of business models followed; one is either doing it themselves or having someone else doing it for them. The generic provider can be called the *Other Provider* (OP).

This is not intended to be facetious. The telecommunications business, of which the Internet is a part, is comprised of those who do it for themselves and those who do it for others. As the software platforms proliferate, so will the number of other providers; therefore, new opportunities will arise. Figure 8-11 illustrates the OP concept.

Selling your own product or selling someone else's product.

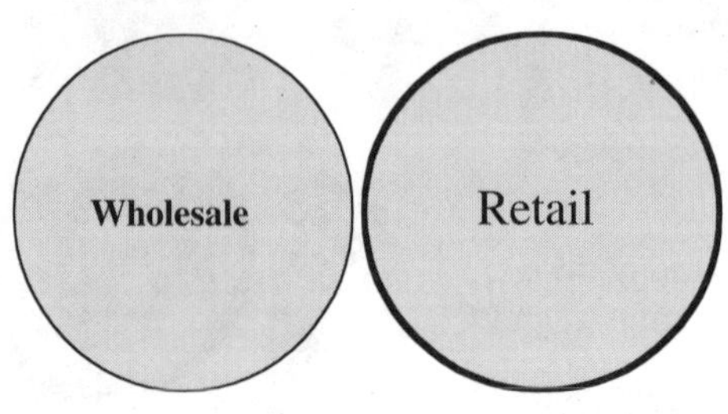

Build it yourself. Build it for someone else and then let someone else brand it with their name.

FIGURE 8-10 The Content Provider

The next generation Internet must able to support information of all kinds.

FIGURE 8-11 The next generation Internet

THE NEXT GENERATION INTERNET: MOBILE INTERNET

Many people have made predictions about the next generation Internet. Some believe it is the next generation wireless network. Others believe it is the next generation wired network. Some people believe that the Internet will not grow beyond simple online shopping. The number of opinions about the future of the Internet coincides with the number of people working in the industry.

The next generation Internet must be one step closer to total market penetration. It must be as commonplace as everyone hopes it will be with a computer and Internet access in every home. The next generation Internet must be accessible from any place. If the next generation Internet is just a faster way of buying collectable baseball cards, the telecommunications industry will have big problems. The Internet has been able to bring the world to places that had no television sets and few, if any telephones. The Internet's biggest challenge is keeping the investment community interested in investing money. The number of failures in 1999 and 2000 is staggering. However, one should consider the failures as typical of the early development of any new business model.

The next generation Internet must be able to do the following:

- Support voice.
- Support video.
- Be WWW compatible.
- Be secure.
- Support broadband data of all kinds.
- Be accessible by all kinds of terminal devices.

- Be ubiquitous throughout the nation.
- Support mobile and wireline Web access.

The key thing to bear in mind is that the next generation Internet will have a mobile component. The aforementioned list represents a list of attributes that the next generation Internet must have. The next generation Internet will be accessible by information brokers and users of all kinds.

ECONOMICS

The government and marketplace are still trying to understand the economics of the Internet. The marketplace has spent a great deal of time making errors, which is a natural process in any early stage industry, while the government has been busy trying to understand how the Internet can generate tax revenues.

The Internet is such a booming business space that it is only a matter of time before the government finds a way of mining tax revenues out of this new business arena. This is why the Advisory Commission on Electronic Commerce was established. This group was established by the United States Congress under the Internet Tax Freedom Act of 1998 to recommend a plan for taxing sales made over the Internet. This is only the beginning. It is fantasy to believe that new accounting rules will not be established for tracking, accounting, depreciating assets, or expensing assets associated with the Internet. Currently, Internet shopping involves a user shopping on a computer; no salespeople are involved, just a computer terminal. This also means fewer jobs in one sector of the economy and hopefully more jobs in a different sector. This can also mean a shift in jobs away from one region to another region of the nation, which means less income tax for one state (if a state income tax exists). The economies of the large industrial powers are now service-oriented and not based on heavy industry. This affects the government's tax base.

The government will continue to examine the way in which revenue is generated on the Internet. Income and other taxes need to be collected in order to pay for the various programs in place and envisioned. By understanding how the Internet is used, the government will likely develop new tax rules to ensure that both state and federal tax structures are not bypassed in any way.

Until 1999, Wall Street was dumping money into any company that had the word "dot-com" in its name. To the casual observer looking in from the outside, the "dot-com" was all the rage, and money was pouring into every idea that had the word "dot-com," even if the company had nothing to do with the Internet. The kinds of questions that should have been asked were as follows:

- What are you selling?
- Why would anyone want to buy it?
- What makes you think they will buy from you?
- How big is the market?
- Are you already operating as a company?
- Are you looking for seed capital or are you further along in the search for financing?
- How much of this market do you expect to capture for your company?
- Why do you think you will be successful?
- What do you plan on doing with this company? Do you plan on taking it public?
- If you plan on going public, when do you expect to do so?
- If you do not plan on taking the company public, what do you plan on doing to increase value to your investor?
- What does your team look like? (In other words, who are your team members?)
- What do you want from me (the investor)?

- What are you bringing to the bargaining table other than a good idea? Do you have any "skin" in the game? "Skin" refers to an individual's personal stake in the business. Usually the stake is one's life savings.
- How much money do you want?
- How much of the company will I get if I invest in you?

Today, the investor has become very "Internet savvy" by asking all of these questions. The challenge for the startup is making sure the business idea becomes a reality.

Wall Street has returned to a level of normalcy, exploring new opportunities at a pace that is less frenetic than it was in 1999. However, this does not mean that no investment opportunities exist. It means the investors are going to take a closer look at the business plan and the people executing that plan.

SUMMARY

The growth of the Internet is based on a number of different factors:

- Technology evolution
- Technology standardization
- New business ideas
- Freedom of creativity on the technical and business fronts
- Market demand

Many of the Internet failures have occurred due to poor execution, not a lack of good ideas. The use of traditional business tools will support the creation of new opportunities and new companies. The Internet is not a mystical object that cannot be understood. Early in the Internet's development, the investment community jumped on to every idea that had the word "dot-com" associated with it. The frenzy for new opportunities had been so great that every rule in the investment book was ignored. Plenty of good ideas were available, but an equal number of people existed who could not execute those ideas. It took a major downturn on the stock market to wake up the investment community.

The primary driver for the Internet's growth will be business/commercial-based. Unless services for which customers want to spend money are available, the Internet will be nothing more than a library and an e-mail service opportunity. The evolution of the Internet has taken a turn into the world of wireless. The growth of the Internet depends on its ability to be used in the two major mediums of telecommunications: wireline and wireless. Unless the Internet can exist in the wireline and wireless worlds, its future will be limited. E-commerce and m-commerce are dependent on one another.

The most important thing for the Internet player or investor to understand is to not lose sight of the basics of doing business. Huckseters of the Internet have used this period of Internet growth and development to create investment opportunities and trouble. New lexicons have been created to describe established "language." Mobile commerce runs the risk of going through the same hype-building that can overshadow good judgment.

Between 1998 and 2000 wireless Internet began to show up on commercial radar screens. Prior to 1998, wireless Internet was purely a standards arena topic, however, it was called wireless data.

Some may take the attitude that "If you build it (bandwidth auctions), they will come and find a billion uses for it." Others take a more pragmatic approach and have the attitude that says, "If you show me the money (for example, show me how this will make money), I will find lots of folks willing to throw billions of dollars into it in order to mind the value." A technical issue can be worked to death in standards or not worked enough in standards. M-commerce is at that point in which the industry is working to transition the technology into the business space as quickly as possible without losing that valuable technical integrity and sacrificing a commercial windfall.

The success of m-commerce is not in question. However, what is in question is: When will m-commerce take off?

APPENDIX A

ACRONYMS, DEFINITIONS, AND TERMINOLOGY

A

AAA server A server that provides authentication, authorization, and accounting security services.

Above the fold The top portion of a Web page that can be seen without scrolling.

Acceptable use policy A policy established for a computing resource (for example, a Web site) that describes how users may access and use the facility.

Access network A portion of a public switched network that connects access nodes to individual subscribers.

Access profile The information maintained by an AAA server for each user. It includes information needed for authentication and accounting. It also specifies the user's access privileges.

Access Tandem (AT) An EC switching system that provides a traffic concentration and distribution function for inter-exchange traffic originating or terminating within an access service area.

Active attack An attempt to break security by altering information. An active attack may or may not require decrypting the information attacked. See passive attack and replay attack.

Ad See advertisement.

Ad serving system A system that inserts ads into Web pages when a user accesses the pages. Different companies may deliver the Web page and the ad using geographically separated servers. The system requires that the designated Web pages have links embedded in them that correspond to certain sizes of ads and categories of products or services. The company delivering the Web page will often receive compensation for attracting users who view the advertising. This same company will have little or no control or interest in the specific advertising messages delivered. Often the company operating the ad serving system is not advertising any of its own products or services, but is acting as a middleman between the advertiser and the company that supplies the Web pages.

Advertisement A message from a company (the advertiser) to potential customers (market) that attempts to influence or reinforce the customers' attitudes and/or behavior toward purchasing the advertiser's products or services, or toward obtaining more information, including further marketing messages, from the advertiser.

Advertising The planning for, development of, and placement of advertisements. Also, the set of advertisements placed by a merchant in a given campaign.

Advisory Commission on Electronic Commerce A group set up by the U.S. Congress under the Internet Tax Freedom Act of 1998 to recommend a plan for taxing sales made over the Internet.

Ad auction A service that enables Web publishers to sell previously unsold ad space at the last minute for whatever price the market will bear. Advertisers are able to buy space at exceptionally low cost-per-thousand. (See also advertisement.)

Ad network A company that contracts with a number of Web sites to sell advertising space on the sites. Some ASPs are nothing more than ad networks.

American National Standards Institute (ANSI) A nonprofit organization that coordinates voluntary standards activities in the United States. The institute represents the United States in two major telecommunication organizations: the *International Standards Organization* (ISO) and the *International Electrotechnical Commission* (IEC).

American Standard Code for Information Interchange (ASCII) A widely accepted standard for data communications that uses a seven-bit digital character code to represent text and numeric characters. When companies use ASCII as a standard, they are able to transfer text messages between computers and display devices regardless of the device manufacturer.

Analog signal A signal that is modified in a constant fashion, such as voice or data.

Applet An application written in Java (see Java) and inserted in an HTML program.

Application Service Provider (ASP) A company that hosts an application and data for one or more customers, providing the hardware, software, infrastructure, and basic maintenance. The service provider supports remote access to the application by the customer, usually over the Internet. The provider usually has expertise in the specific application hosted and may provide enhancements to it. Advantages to the customer versus running the application in-house include guaranteed performance, a robust data center, in-depth support from a skilled staff, and better economics.

ASP See Application Service Provider

Asymmetric Digital Subscriber Line (ADSL) Modems attached to twisted pair copper wiring that transmit from 1.5 Mbps to 9 Mbps downstream (to the subscriber) and from 16 Kbps to 800 Kbps upstream, depending on the line distance.

Asynchronous Transfer Mode (ATM) A multiplexing and switching technique that organizes information into 53 byte cells. Each cell of data is transmitted asynchronously.

Audit trail Information on the major steps in the progress of a transaction that allows an auditor to determine that no errors appear in the reporting of the transaction. In electronic commerce, an audit trail can help resolve programming errors and discrepancies in how a transaction is recorded by the parties to the transaction.

Authentication A process where information is exchanged between a communications device (typically a mobile phone) and a communications network that enables the carrier or network operator to confirm the true identity of the unit. This inhibits fraudulent use of the mobile unit. Another way of viewing authentication: Authentication is the process of user identity confirmation. Identity confirmation can involve checking handset/terminal device identity by interpreting "secret" keys/data messages. If the data keys/data messages have been altered or do not show specific format, the call will not be completed.

B

B2B Business-to-Business marketing and commerce.

B2SB Business-to-Small Business marketing and commerce.

B2C Business-to-Consumer marketing and commerce.

Backbone A common distribution channel that carries analog or digital telecommunications signals for many users. Also, the central distribution cable from an interface.

Bandwidth The width of a radio channel (in Hertz) that can be modulated to transfer information.

Banner A graphic element on a Web page used to title the page, head a new section, present a company's or advertiser's message, or provide a link to another page.

Basic Rate Interface (BRI) In ISDN, the network interface that provides 144 Kbps information transfer to the consumer premises on twisted pair as defined in ANSI Standard T1.607.

Bearer In the communications industry, this term refers to a transmission channel that is used to carry data. In ISDN, 64 Kbps bearer channels are used. These channels are used to carry data.

Beta software Software that has been developed to the point where it can be tested for reliability and usability in customer environments, but not to the point where it can be commercially sold, used, or supported. Beta sites are customer sites where the software is tested. Beta customers are often given some form of preferential treatment or "Most Favored Nation" status.

Bill presentation The delivery of a bill (invoice) to a customer for payment. Third party specialists compete with banks to provide electronic bill presentment over the Internet, especially for companies that serve large numbers of consumers (that is, utility companies).

Billing system A system that records the occurrence of a call or some event, the identity of the originating party, the identity of the destination party, and the time length of the call. This system must also process the data for the rendering of a bill to the subscriber.

Bit The smallest part of a digital signal, typically called a data bit. A bit usually assumes two levels: either a zero (0) or a one (1).

Bit error rate (BER) A measurement used to determine the quality of a digital transmission channel. BER measures the ratio of bits received in error compared to the total number of bits transmitted.

Brand A name or symbol chosen by a company to uniquely identify a set of products and/or services that the company has established or hopes to establish in the minds of its customers. The brand name is meant to serve as an indication of quality and trustworthiness. It is usually legally registered as a trademark with a governmental organization to prevent other companies from using it. A brand is usually established in customer's minds through marketing, particularly advertising, and by satisfying customer experiences with the product or service. A brand is valuable to a company in encouraging repeat purchases and extending the brand, with its aura of quality and trustworthiness, to new products and services. Examples of widely known and successful brands include Coca-Cola (a trademark of the Coca-Cola Company) and Amazon (a trademark of Amazon, Inc.). Branding (the establishment of a brand) is generally judged to be extremely important in electronic commerce because of the ease with which customers can find, compare, and do business with a wide variety of competitive companies offering similar products and services on the Internet.

Brick and mortar A traditional business operation that commonly deals with its customers face-to-face in an office or storefront that the business owns or rents (for examples, a retail store or a bank branch). Internet businesses usually have lower costs and greater flexibility than brick and mortar operations.

Bridge A data communications device that connects two or more networks and transmits information between the networks.

Broadband Typically refers to voice, data, and/or video communications at rates greater than wideband communications rates (1.544 Mbps).

Broadband Integrated Services Digital Network (B-ISDN) A digital network with ATM switching operating at data rates in excess of 1.544 or 2.048 Mbps. ATM enables the transporting and switching of voice, data, image, and video over the same infrastructure.

Broker An individual or company that is an intermediary between a seller and a buyer. A stockbroker specializes in stocks, bonds, and other investments.

Button A small advertisement that can provide a constant presence on a Web site although costing considerably less than a banner. Buttons can build brand awareness and attract traffic to the sponsor's Web site, but their message must be limited and simple.

Bursty data Describes data rates that fluctuate widely with no predictable pattern.

Busy hour A time-consistent hour in a specific measurement period when the total load offered to a group of trunks, a network of trunks, or a switching system is greater than at any other time-consistent hour during the same measurement period.

C

Call processing Steps that occur during the duration of a call. These steps are typically associated with the routing and control of the call.

Call routing In circuit switching, the process of determining the path of a call from the point of origination to point of destination.

Cache High-speed memory that stores data for relatively short periods of time under computer control in order to speed data transmission or processing. Web pages that are used frequently are often cached close to users so that they can be delivered quickly without requiring repeated transmissions over the entire network between the Web server and the client machine.

CCS (One Hundred Call Seconds) A measurement of telephone usage traffic used to express the average number of calls in progress or the average number of devices used.

CDR See Call Detail Reporting

Cell site A transmitter-receiver tower operated by a wireless carrier (typically cellular or PCS) through which radio links are established between a wireless system and mobile and portable units.

Central Office A term to describe local switches used by the local telephone companies. The term is also synonymous with end office.

CLEC See competitive local exchange carrier.

Click on the Web The act of selecting and following a link by placing one's mouse cursor on the text, graphics, banner, or button identifying the link and depressing a button on the mouse.

Click-through rate The percent of individuals viewing a Web page who click on a specific banner ad appearing on the page. Overall, click-through rates have declined substantially since the early days of electronic commerce on the Web.

Click stream The Web sites and Web pages a user visits, either in a single online session or over a longer period of time. Information about a user's click stream on a single site and some information about the click stream over multiple sites can be collected by specialized organizations using cookies. The information can be used to improve Web sites, customize Web sites, and target appropriate marketing messages from ad-serving systems.

Closed loop reporting The ability to measure the effectiveness of a particular ad on the Web by tracking which ad viewers actually bought which product, requested a catalog, or took other specific actions on the Web site.

Co-branding Loosely, the appearance of two company names on a Web page or Web site, implying that a product or service is provided jointly by the two companies. Often the site belongs to a company with a significant customer base, market awareness, or marketing power although the other company on the page or site is actually providing the bulk of the product or service. In another variety of co-branding, an advertiser provides information about its offering in the midst of ostensibly objective information on the site.

Community In the context of the Internet and electronic commerce, people who participate in an online discussion group or bulletin board or who return frequently to a Web site because of a common interest in a given subject. One business strategy developing on the Internet is to create a community and then sell access to the group for marketing purposes. (See bulletin board system in the hard copy dictionary.)

Common Channel Signaling (CCS) Also known as "Out-of-Band" signaling. Describes a scheme where the content of the call is separated from the information used to set up the call (signaling information).

Competitive Access Provider (CAP) A company that competes with local telephone companies to provide access from the customer's premise to long-distance carriers.

Competitive Local Exchange Carrier (CLEC) A description of the new competitive carriers (as encouraged by the Telecommunications Act of 1996) that are competing in the local loop marketplace.

Control signals Control signals are used for special auxiliary functions that are beyond a service provider's network. These signals communicate information that enables or disables certain types of calls. One example would be call barring.

Convergence Addresses the technical and business aspects of integration of technology and business.

Cookie A capability of some Web browsers that enables Web servers to store information about user visits to the Web site on the hard disk in the user's PC or workstation. Because it can be used to identify repeat visitors, the cookie enables on the fly customization of a Web site to feature items the user showed an interest in during previous visits. The cookie also enables a Web server to track the sequence of a session on a Web site, including how long the user spent on each Web page. Although a boon to marketing on the Web, the cookie raises some privacy issues because it removes some of the traditional anonymity associated with viewing Web sites and uses a small portion of the user's hard disk.

Cost-per-click (CPC) A method of charging for banner advertisements on the Web on the basis of how many viewers click on the ad rather than charging on a flat rate basis for the advertising space. The CPC model is generally considered to favor advertisers rather than publishers because it ignores the brand building value of an ad that is seen, but not clicked on. Generally, it gives the publisher no control over the content or appearance of the ad and may require the publisher to allocate space to poorly performing ads until the contracted number of clicks have been achieved.

Cost-per-thousand (CPM) The cost in any given media for a thousand impressions. On the Web, the CPM for impressions on a Web site that attracts an undifferentiated variety of consumers (for example, a typical search engine) is on the order of $5. The CPM for advertising on a Web site with a specific audience that is likely to make sizable purchases (for example, corporate telecommunications managers) can be in the neighborhood of $100.

Country code A one-, two-, or three-digit number that identifies a country or numbering plan to which international calls are routed. The first digit is always a world zone number. Additional digits define a specific geographic area, usually a specific country.

CPC See cost-per-click.

CPM See cost-per-thousand.

Customer care system A customer profile database system used to support customer complaints, add new subscribers, remove subscribers, and contain customer profile information.

Customer premises equipment (CPE) All telecommunications terminal equipment located on the customer's premises, including telephone sets, private branch exchanges (PBXs), data terminals, and customer-owned, coin-operated telephones.

D

Data compression A technique for encoding information so that fewer data bits of information are required to represent a given amount of data. Compression enables the transmission of more data over a given amount of time and circuit capacity. It also reduces the amount of memory required for data storage.

Data protection The prevention of the passing of an individual's personal information from one computer system where the information legitimately resides to other computer systems without the consent of the individual. In the United Kingdom, the Data Protection Act of 1984 requires the registration of every data user who processes personal information.

Data terminal equipment (DTE) In a data communications network, the data source such as a computer, and the data sink such as an optical storage device.

Data warehouse An information management service that stores, analyzes, and processes information that is derived from transaction systems.

Database A collection of interrelated data stored in computer memory with a minimum of redundancy. Database information held in a computer-accessed memory usually is subdivided into pages, with each page accessible to all users unless it belongs to a closed user group.

Demographics In marketing, the description of a group of consumers, most often age, gender, household income, level of education, marital status, employment status, number of people in the household, and region of the country. More loosely, a description of any group of customers whether consumers or businesses.

Descriptive billing A system of credit card billing in which individual charge slips are not returned with the bill. Federal Reserve Board Regulation Z requires that if a charge slip is not enclosed, the bill must provide the date, the amount, the seller's name, and the location where the transaction took place for each transaction.

Diffie-Hellman A public key algorithm that can be used for deriving and distributing private keys over insecure networks. Diffie-Hellman itself cannot be used to encrypt or decrypt messages. The patent on the algorithm expired on April 29, 1997.

Digital signal A signal that has a limited number of discrete states, usually two. In contrast, an analog signal varies continuously and has an infinite number of states.

Digital Signal 0 (DS0) A 64-Kbps digital be it voice or data.

Digital Signal 1 (DS1) Twenty-four voice channels packed into a 193-bit frame and transmitted at 1.544 Mbps. The unframed version, or payload, is 192 bits at a rate of 1.536 Mbps.

Digital Signal 2 (DS2) Four T1 frames packed into a higher level frame transmitted at 6.312 Mbps.

Digital Signal 3 (DS3) Twenty-eight T1 frames packed into a level frame transmitted at 44.736 Mbps.

Digital subscriber line (DSL) A two-wire, full-duplex transmission system that transports user data between a customer's premises and a digital switching system or remote terminal at 144 Kbps.

Directory A structure for organizing files on a computer or network. Directories may be protected or shared using various methods for information security.

Domain name The unique name that identifies an Internet site.

DSL See Digital Subscriber Line.

DVD A technology that uses five-inch discs similar to CD-ROMs, but can hold much more data. A DVD disc can deliver multichannel sound and TV images three times sharper than videotape. DVD players, expected to debut in late 1996, will be interoperable with televisions, stereos, personal computers, and some video games. Unlike CDs, DVDs can hold data on both sides and can provide either one or two layers of data on each side. A single-sided, single layer DVD holds 4.7 gigabytes of data while a double-sided, double layer DVD holds 17 gigabytes.

E

Economic profit Return on investment in excess of the average cost of capital in the overall market. Alternatively, profits excluding profits from the increase in value of inventory that the company holds.

Electronic mail (E-mail) Messages, usually text, sent from one person to another via computer.

Electronic tax filing An offering of the *Internal Revenue Service* (IRS) that enables taxpayers to file their tax returns on diskette or over a network. In 1994, 11.8 million tax returns were filed electronically or about 11 percent of all individual tax returns. The number was down 15.8 percent from 1993 because of increased IRS efforts to prevent fraud, which can be facilitated when tax returns are filed electronically. Fraud has caused the IRS to question its goal of getting 80 percent of Americans to file electronically by the year 2001.

Element The basic building block of an HTML document. When an element consists of a start tag, end tag, and text or some other content, it can also be referred to as a "container." An "empty element," such as that commanding a line break, has only one tag and no content.

Encryption A process of a protecting voice or data information from being obtained by unauthorized users. Encryption involves the use of a data processing algorithm (formula program) that uses one or more secret keys that both the sender and receiver of the information use to encrypt and decrypt the information. Without the encryption algorithm and key, unauthorized listeners cannot decode the message.

End office (EO) An EC switching system that terminates station loops and connects the loops to each other and to trunks.

Engineered capacity The highest load level at which service objectives are met for a trunk group or a switching system.

Erlang Amount of voice connection time with reference to one hour. This is used to describe the aggregate traffic flowing over the transmission facility. For example, a six minute call is .1 Erlang.

Ethernet A transmission protocol for packet-switched *local area networks* (LANs). Ethernet is a registered trademark of Xerox Corporation.

Exchange carrier A telephone company, generally regulated by a state regulatory body, that provides local (intraLATA) telecommunications services.

Extensible Markup Language (XML) A document description language, primarily used to define Web pages and applications that are compatible with *Hypertext Markup Language* (HTML), but provide more capabilities and flexibility. XML removes presentation constraints imposed by HTML and does not necessarily require a Web browser. Although both languages are platform independent, XML is also database and file format independent. To display an XML document (or run the XML application), the user's machine must be equipped with an XML parser. For basic XML documents, any XML parser can infer the document rules by the way the elements are presented. For more complicated documents or applications, the author must develop or utilize a *document-type definition* (DTD) that defines the structure and rules for a given type of document. The XML parser refers to the DTD in order to determine how to display the document or run the application. More or less standard DTD's have been developed for engineering, law, medicine, and other areas of knowledge.

F-G

Facilities The transmission parts (elements) of a service provider. Sometimes used in more general terms to describe buildings and utilities.

Firewall A term that refers to a physical and electronic method of protecting a computer from outside attack. This protection involves hardware and software.

Flat rate Price setting principles for a service provider that wishes to charge the same fee for calls regardless of the number of calls made or the duration of each call. Usually, flat rates are single monthly or periodic charges.

First-time visitor A visitor of a Web site who has never visited the site before.

Gateway A device or facility that enables information to be exchanged between two dissimilar computer systems or data networks. A gateway reformats data and protocols in such a way that the two systems or networks can communicate.

Grade of Service (GOS) An estimate of customer satisfaction with a particular aspect of service. It also refers to the probability that a call will fail due to the unavailability of links or circuits.

H

Handoff The process of reassigning subscriber handsets to specific radio channels as the handsets move from cell site to cell site.

Hard copy A physical document, usually text on paper.

Hard handoff A "break-before-make" form of call handoff between radio channels. In this scenario, the mobile handset temporarily (time measured in milliseconds) disconnects from the network as it changes channels. The radio protocols, AMPS, TDMA, and GSM, support only hard handoff.

Header control Information inserted in front of data when the data is encapsulated for network transmission.

Helper A program used by a WWW browser to process files that the browser itself cannot, for example, a sound file, JPEG image, MPEG movie, or a compressed file. Also known as a viewer.

High data rate Digital Subscribe Line (HDSL) Modems on either end of one or more twisted pair wires that deliver T1 or E1 speeds. At present, T1 requires two lines and E1 requires three. (See SDSL for one line HDSL.)

High Usage Trunk Group A transmission facility used only for routing large volumes of traffic to a single point or set of points.

I

Incumbent Local Exchange Carrier (ILEC) A telephone service carrier that was operating in the local loop market prior to the Telecommunications Act of 1996 and the divestiture of the AT&T bell system. In other words, it was the local telephone company most people grew up with.

Information service The offering of a capability for generating, acquiring, storing, transforming, processing, retrieving, utilizing, or making information available via telecommunications. This includes electronic publishing, but does not include any use of any such capability for the management, control, or operation of a telecommunications system or the management of a telecommunications service.

Inline image An image that appears in the midst of text on a Web page.

Integrated Services Digital Network (ISDN) A structured all digital telephone network system that was developed to replace (upgrade) existing analog telephone networks. The ISDN network supports for advanced telecommunications services and defined universal standard interfaces that are used in wireless and wired communications systems.

Interconnection The connection of telephone equipment or communications systems to the facilities of another network. The FCC regulates interconnection of systems to the public switched telephone network. (See also: bypass.)

Inter-exchange carrier (IXC) A carrier company in the United States, including Puerto Rico and the Virgin Islands, that is engaged in the provision of interLATA, interstate, and/or international telecommunications over its own transmission facilities or facilities provided by other inter-exchange carriers.

International Carrier (INC) A carrier authorized to provide inter-exchange communications services outside World Zone 1 using the international dialing plan; however, the carrier has the option of providing service to World Zone 1 points outside the 48 contiguous states of the United States.

International Gateway Facilities (IGFs) Transmission facilities used by International Gateway Switches. An International Gateway Switch is used by the long-distance (inter-exchange carrier) to interface with international telecommunications networks.

International Telecommunications Union (ITU) A European telecommunication standards body. A counterpart of ANSI.

Internet The major network running the Internet protocol across the United States and Canada. The Internet consists of more than 30,000 hosts and includes sites at universities, research laboratories, corporations, and nonprofit agencies.

Internet Advertising Bureau (IAB) An association of companies that advertise on the Internet. Key activities include establishing terminology and standards for measuring advertising, conducting research relevant to electronic commerce and advertising, and addressing industry issues such as privacy and taxation.

Internet Service Provider (ISP) A vendor that provides access to the Internet and the World Wide Web.

Internet Tax Freedom Act of 1998 A law passed by the U.S. Congress that placed a three-year moratorium on new taxes on Internet access fees and prohibited multiple and discriminatory taxes on electronic commerce. The Internet appears likely to drive major simplification of the sales tax laws, which vary greatly between different states and local governments across the country. The act also set up the Advisory Commission on Electronic Commerce.

Intranet A private network inside a company or organization that not only serves as the company's information network, but also access to the public Internet. The Intranet appears like another server to the Internet.

IP telephony Technology that supports voice, data, and video (both broadcast and interactive) over IP-based LANs, and WANs and over the Internet.

IP service The carrying of data traffic using the Internet protocol for a fee. This is the essential service provided by Internet Service Providers and Network Service Providers.

IP-SEC An IETF standard for encrypted sessions between corporate firewalls.

J

Java A programming language developed by Sun Microsystems to support widespread software distribution, particularly over the Web. It is a slimmed down and more secure version of the C++ programming language. Java was originally developed for use with set-top boxes (see set-top boxes in the hard copy dictionary). Java runs on Web servers and downloads small application programs called applets (see applets) to Web browsers on an as needed basis. Because of strict built-in controls over software distribution, the Java design protects against the delivery of incompatible software or viruses.

JavaBeans A software component model associated with the Java programming language. A JavaBean component is an object that can be used repeatedly by visually manipulating it into larger solutions. JavaBeans can range from low-level components such as a scroll bar to complete products like Corel's Office for Java.

JavaScript A cross-platform, Web-based scripting language that will enable a layperson to author HTML pages that use Java applets, objects, and classes without having to know how to program in Java. It can also be used to build stand-alone applications that can run on either clients or servers.

L

Local Access and Transport Area (LATA) As designated by the Modification of Final Judgment, an area in which a local exchange carrier is permitted to provide service. It contains one or more local exchange areas, usually areas with common social, economic, or other interests.

Local area network (LAN) A private network offering high-speed digital communications channels for the connection of computers and related equipment in a limited geographic area. LANs use fiber optic, coaxial, or twisted-pair cables or radio transceivers to transmit signals.

Local exchange carrier (LEC) A company that provides telecommunications service within a *local access and transport area* (LATA).

M

Metadata A description of a set of useful data, usually in digital form and conforming to a well-defined template. Also, an index composed of such descriptions. Analogous to a card in a library card catalogue system, a metadata description of a set of data might include an abstract of the data and the data's format, multimedia content, location, and usage fees.

Metcalfe's Law The total value of a network to its users grows as the square of the total number of users. The law was developed by Bob Metcalfe originally to convince people of the value of Ethernet and has also been referred to by George Gilder as the law of the telecoms. An important consequence of the law is that the ratio of value to cost for adding one more network user grows disproportionately (increasing returns) as the network gets bigger.

N

Navigation The actions of a user exploring a Web site (or multiple Web sites) or searching for information or resources. Also, the design of a Web site, including links, text, graphics, and page layouts to facilitate navigation by users.

Netcasting The broadcasting of information over the Internet using server push technology. Generally, limited to Web casting.

Netizen A citizen of cyberspace. An Internet user.

Nettop An alternate term for Internet appliance (see Internet appliance).

Network A series of points interconnected by communications channels, often on a switched basis. Networks are either common to all users or privately leased by a customer for some specific application.

Network Access Server (NAS) A device providing temporary, on-demand access by individuals to a network. NAS most commonly refers to a remote access server providing dial-up access over analog or ISDN telephone lines.

Network element A facility or the equipment used in the provision of a telecommunications service. The term includes subscriber numbers, databases, signaling systems, and information sufficient for billing and collection or used in the transmission, routing, or other provision of a telecommunications service.

Network interconnection The interconnecting of two more networks to one another.

Network Operations Center (NOC) A center responsible for the surveillance and control of telecommunications traffic flow in a service area.

Network Service Provider As opposed to Internet Service Providers who sell to end-users, Network Service Providers sell high volume Internet backbone capacity to Internet Service Providers. Network Service Providers often sell to end users as well. Sometimes called "carriers," Network Service Providers include MCI, AT&T, Sprint, and European Postal, Telegraph, and Telephone (PTT) agencies.

NSP See Network Service Provider.

O

Open Systems Interconnection (OSI) Model The OSI model is an internationally accepted framework of standards for telecommunications between different systems made by different vendors. The model organizes the telecommunications functions into seven different layers. This model enables engineers to isolate and classify telecommunications into discrete functions or activities.

Opt-in The expressed act by a customer of granting a marketer permission to deliver marketing messages to the customer. The permission is typically granted by registering on a Web site or responding to an unsolicited e-mail. The messages are typically delivered by e-mail.

Opt-out The instruction by a customer to a marketer to halt the delivery of marketing messages, typically periodic e-mails. Most responsible marketers embed opt-out procedures in their marketing messages and make it easy for customers to opt-out.

P

Packet switching A mode of data transmission in which messages are broken into increments, or packets, each of which can be routed separately from a source and reassembled in the proper order at the destination.

Page view A unit for measuring Web site readership that corresponds to one person viewing one page, or at least a portion of a page, one time. If a person leaves a certain page and returns to it in the same visit to the Web site, two page views are counted. Page views are more difficult to measure than hits, but are a much more useful and relevant measure of the attention a page garners from viewers.

Paperwork Reduction Act of 1995 An act that requires all federal agencies to use electronic filing forms to reduce demands on the public for paperwork 10 percent by 1996-97 and five percent annually through the year 2000.

Partial URL A Uniform Resource Locator that refers to locations in relation to the document currently being viewed, typically a location on the same server or Web site.

Passive attack An attempt to break security by capturing information without altering it. See active attack, one-time password system, and replay attack.

Path minimization (efficient routing of a mobile call) Path minimization is the process of efficient fixed network (the non-wireless portion of the call) routing of wireless calls tables in the wireless carrier switches.

Pay-per-click The ability to pay, typically in small amounts or micro-payments (see hardcopy dictionary) for information or entertainment received over the Internet or the Web. The term implies that each payment is automatic and that simply requesting the information provides agreement to pay for it.

Plug-in Software that runs on a client processor and in conjunction with a Web browser to provide specialized manipulation of data obtained over the network. Plug-ins and ActiveX components have the advantages over Java applets because they run faster and do not have to be downloaded because they already exist on the client machine. Plug-in and ActiveX disadvantages from applets are that they must be manually acquired and installed by the user before they can be used.

Point of Interface (POI) The physical location marking the point at which the local exchange carrier's service ends.

Point of presence A physical location established by an interexchange carrier within a LATA for the purpose of gaining LATA access. The point of presence is usually a building that houses switching and/or transmission equipment, as well as the point of termination. (See also: point of termination.)

Primary rate interface In the *Integrated Services Digital Network* (ISDN), a channel that provides digital transmission capacity of up to 1.536 Mbps (1.984 Mbps in Europe) in each direction. The interface supports combinations of one 64-Kbps D channel and several Kbps B channels, or H channel combinations. PRI is the inter-switch component of ISDN and is the compliment of BRI.

Private branch exchange (PBX) A private switching system serving an organization, business, company, or agency, and usually located on a customer's premises.

Protocol (1 - rules) A precise set of rules and a syntax that govern the accurate transfer of information. (2 - connection) A procedure for connecting to a communications system to establish, carry out, and terminate communications.

Protocol conversion The translation of the protocols of one system to those of another to enable different types of equipment, such as data terminals and computers, to communicate.

Provisioning The operations necessary to respond to service orders, trunk orders, and special-service circuit orders, and to provide the logical and physical resources necessary to fill those orders.

R

Registration A Web site visitor's input of personal or business information into a form provided on the Web site. Registration enables the Web site owner to better understand viewers and customers, increases the likelihood of users returning to the site, and usually gains permission to send marketing messages to the visitor. The Web site owner may provide the registrant with access to restricted information (for example, a research report), a Web site customized for the user, a free service (for example, free e-mail), a discount on products or services, entry in a sweepstakes, or some other enticement to register. Users generally expect and Web sites generally provide a privacy statement describing how the information will be used.

Registration service A generally free service on the Internet that requires registration by the user, sometimes including detailed user information, in return for access to the service.

Ring network A data network of circular topology in which each node is connected to its neighbor to form an unbroken ring. A ring network in which one of the nodes exercises central control often is called a loop.

Router See routing switcher.

Routing switcher (1 - general) An electronic device that connects a user supplied signal (audio, video, and/or data) from any input to any user selected output. Inputs are called sources. Outputs are called destinations. (2 - network) A device that forwards packets of a specific protocol type (such as 'P) from one logical network to another. These logical networks can be the same type or different types. A router receives physical layer signals from a network, performs data link and network layer protocol processing, then sends the information to the next network node.

S

Search engine On the World Wide Web, a Web site that catalogues a vast number of Web pages and other documents on the Internet, and provides links to them and descriptions of them for users. The search engine provides a list of hopefully relevant documents and Web pages in response to queries by users, usually in the form of words or phrases that indicate the topic in which the user is interested. Search engines are often the key element in a portal.

Security gateway A system that acts as the communications gateway between external, untrusted systems and trusted hosts on their own sub-network. It provides security services for the trusted hosts when they communicate with external untrusted systems. When a security gateway is providing services on behalf of one or more hosts on a trusted sub-net, the security gateway establishes the Security Association on behalf of its trusted host and provides security services between the security gateway and the external systems. In this case, the gateway uses the IP Authentication Header, while all of the systems behind the gateway on the trusted sub-net may take advantage of IP Authentication Header services without having to implement them.

Server (1 - telecommunications network) The equipment or a call carrying path that responds to a customer's attempt to use a network. (2- LAN) A processor that serves users on a local area network, for example, by storing and managing data files or by connecting users to an external network.

Service provider A generic name given to a company or organization that provides telecommunications service to customers (subscribers). (See also Network Provider and Reseller.)

Signaling System Number 7 (SS7) An out-of-band common-channel signaling protocol standard that is designed to be used over a variety of digital telecommunication switching networks. It is optimized to provide a reliable means for information transfer for call control, remote network management, and maintenance.

Soft handoff The reverse of the hard handoff scenario. Soft handoff is a "make-before-break" form of call handoff between radio channels, whereby the mobile handset temporarily communicates with both the serving cell site and the targeted cell site (one or more cell sites can be targeted) before being directed to release all but the final target cell site radio channel. Currently, only CDMA supports soft handoff.

Star network A data network with a radial topology in which a central control node is the point to which all other nodes join.

Switching (1 - general) The process of making and breaking (connecting and disconnecting) two or more electric circuits. (2 - telecommunications) The process of connecting appropriate lines and trunks to form a communications path between two or more stations. Functions include transmission, reception, monitoring, routing, and testing.

Synchronous (1 - general) In step or in phase, as applied to two or more devices: a system in which all events occur in a predetermined timed sequence. (2 - data communications) An operation in which all involved parties are driven by a single master clock. All actions begin and end in compliance with the beat of this master clock.

T

Tag A command in HTML, enclosed within the lesser than and greater than signs (< . . . >), which marks specified text as a heading, paragraph, list, and so on, and used for including images, forms that accept user input and hypertext links within a document. Usually, a start tag and end tag are paired around the content they refer to and form a "container."

Tandem switch A switch that supports a network topology where connectivity between locations is attained by linking several locations together through a single point. The tandem switch is similar to the "traffic cop": directing traffic from several streets into other groups of streets. Tandems are similar to gateways.

Targeting The design of a Web site or of a marketing campaign (primarily the selection of Web sites in which advertisements are placed, but also the content of the advertisements) to reach and appeal to certain classes or types of customers.

Telecommunications The transmission between or among points specified by the user of information of the user's choosing (including voice, data, image, graphics, and video), without change in the form or content of the information.

Telecommunications carrier Any provider of telecommunications services. A telecommunications carrier shall be treated as a common carrier under this Act only to the extent that it is engaged in providing telecommunications services.

Telecommunications service The offering of telecommunications for a fee directly to the public or to such classes of users as to be effectively available directly to the public, regardless of the facilities used.

Terminal equipment Refers to the computers, telephones, and other data or voice devices at the end of a telephone line.

Terminals Devices that typically provide the interface between the telecommunications system and the user. Terminals may be fixed (stationary) or mobile (portable).

Traffic Communication over a network and especially the amount of communication over a network. In network technology, traffic is usually measured in bits per second or packets per second. In Web-based marketing, traffic refers to the number of customer visits to a site or page and is measured in a variety of ways.

Traffic engineering planning An activity that determines the number and type of communication paths required between switching points and the call processing capacity of the switching equipment.

Translation (1 - general) The conversion of information from one form to another. (2 - switching system) The conversion of all or part of a telephone address destination code to routing instructions or routing digits.

Transmission Control Protocol/Internet Protocol (TCP/IP) The protocol that manages the bundling of outgoing data into packets, manages the transmission of packets on a network, and checks the bundles for errors. This protocol is a networking protocol that supports communication across interconnected information networks, that is, the Internet.

Tunneling An architecture that enables the point-to-point transmission of electronic data in a format different from the format in which it originated and was ultimately received. Tunneling can be used to send non-IP protocols (for example, IPX or AppleTalk) over an IP network such as the Internet. It can also be used to transmit unencrypted data over the transmission media in encrypted format without requiring encryption by the host or application at either end. The Point-to-Point Tunneling Protocol encapsulates the Point-to-Point Protocol for control purposes.

Type 1 A connection that is a trunk-side connection to an end office. The end office uses a trunk-side signaling protocol in conjunction with a feature known as *Trunk with Line Treatment* (TWLT).

Type 1 with ISDN An ANSI standard ISDN line between the LEC end office and the MSC. The ISDN connection should be capable of providing connectivity to the PSTN that will support ISDN *Primary Rate Interface* (PRI) and ISDN *Basic Rate Interface* (BRI). The ISDN connection is a variation of the Type 1 connection.

Type 2A A trunk-side connection to the LEC's access tandem. This connection enables the MSC to interface with the access tandem as if it were a LEC end office. This connection enables the wireless carrier's subscribers to obtain presubscription. The service provider will have access to any set of numbers within the LEC network.

Type 2B A similar inter-connection type to the high usage trunk groups established by the LEC for its own internal routing purposes. In the case of wireless carrier interconnection, the Type 2B should be used in conjunction with the Type 2A. When a Type 2B is used, the first choice of routing is through a Type 2B with overflow through the Type 2A.

Type 2C A telephone system inter-connection that is intended to support interconnection to a public safety agency via a LEC E911 tandem or local tandem. This connection enables a wireless carrier to route calls through the PSTN to the *Public Safety Answering Point* (PSAP). This connection supports a limited capability to transport location coordinate information to the PSAP via the LEC.

Type 2D A telephone system interconnection that is intended to support interconnection to the LEC's operator service position. This connection enables the LEC to obtain *Automatic Number Identification* (ANI) information about the calling wireless subscriber in order to create a billing record.

Type S A telephone system interconnection that only carries control messages. The Type S is a SS7 signaling link from the wireless carrier to the LEC. The Type S supports call setup via the *ISDN User Part* (ISUP) portion of the SS7 signaling protocol & TCAP querying. The Type S is used in conjunction with the Type 2A, 2B, and 2D.

U–X

Uniform Resource Location (URL) This is a naming model that gives the servers and content names, tells where the various items are located, and tells how to get the items (identifies the protocol needed to get the items).

Validation Often confused with authentication. Authentication essentially certifies the user as an either "real" or "fake" and also as either "good" or "bad", whereas validation certifies the "permission" to complete the call. For example, a user calls another party using a mobile handset. The carrier certifies that the handset is the real one and is not a clone being used by an unauthorized person. Once the handset has been given the "thumbs up," the carrier checks to see if the call is even allowed under the user's billing plan. This next step is validation: being given the final "green light" to complete the call.

Video An electrical signal that carries TV picture information. (See also: video signal.)

Video on demand (VOD) A service that telephone companies were seeking to bring to their subscribers. VOD would enable a subscriber to request any video they wished to see.
Visit The act of accessing and viewing a World Wide Web site.

Visitor An individual who views a Web site.

Voice Mail (VM) Provides the subscriber with voice mail services, which include not only the basic voice recording functions, but also time of day recording, time of day announcements, menu driven voice recording functions, and time of day routing.

Voice over IP Technology that supports the transmission of voice over networks running the IP protocol, such as the Internet, with sufficient speed and continuity to permit two-way conversations such as a telephone call. The voice quality is not as good as standard telephone calls placed over the *Public Switched Telephone Network* (PSTN), but the cost of calls using voice over IP is extremely low regardless of distance.

Wideband The passing or processing of a wide range of frequencies. The meaning varies with the context. In an audio system, wideband may mean a band up to 20 kHz wide; in a TV system, the term may refer to a band many megahertz wide.

xDSL A set of large-scale, high-bandwidth data technologies that can use standard twisted-pair copper wire to deliver high speed digital services (up to 52 Mbps).

X.25 The X.25 protocol is a data link layer protocol. Specifically, X.25 uses the LAPB portion of this layer.

APPENDIX B

NETWORK INTERCONNECT

This appendix has been included in order to highlight the interconnection issues surrounding the Internet and other networks. I had briefly touched on the issue of interconnection in the discussion surrounding the wireline and wireless Internet. Interconnection is not just the act of connecting networks together; interconnection is the act of communication between networks. The communication between networks is seamless when the network are interoperable (compatible). The lack of compatibility between the wireline and wireless network resulted in the creation of two primary interconnection documents. These documents define how the two networks are connected and communicate to one another. The levels of communication vary. The desire of the different subscriber bases to communicate with one another drove network interconnection between the two network types. Total interoperability means total seamlessness between the networks. Seamlessness is good for the consumer and ultimately good for the industry.

The Internet will face the same issues unless the compatibility issue is faced head on between the different Internet types. Unless the industry wishes to see separate and distinct Internets form, the only choice is to view the Internet as a network compatibility imperative.

Work on TIA's IS-93 began in 1992 when it became clear that technology development had elevated the wireless switch from the technical functionality of a PBX to that of a switching system equal to that of a telephone company switch. A number of wireless industry players had hoped to use IS-93 as a way of forcing the RBOCs into renegotiating their interconnection agreements. The most significant difference between IS-93 and TR-NPL-000145 Issues 1 and 2 was the acceptance of a bi-directional signaling relationship between the cellular carriers and the PSTN. By acknowledging that the SS7 signaling occurred in two directions, the RBOCs could interact with the cellular carrier as a co-carrier, and therefore, the cellular carriers could force acceptance of mutual compensation. GR-145-CORE was written (by Bellcore) in response to IS-93 Revision A. Bellcore's GR is a proprietary product sold by Bellcore (now known as Telcordia).

The reason the documents were written is really not of any consequence. The relevant point is that both documents support the interconnection business.

Both documents were written to support technical and business perspectives and objectives of their respective interest groups: GR-145-CORE supports the objectives of the RBOCs and IS-93 supports the objectives of the wireless industry (specifically, the *Cellular Telecommunications Industry Association* (CTIA) and the *Telecommunications Industry Association* (TIA). IS-93 is intended to support the entire wireless community. IS-93 is a standard. GR-145-CORE is a company proprietary document sold for profit.

GR-145-CORE is service- and implementation-oriented. In another words, GR-145-CORE interconnection descriptions provide more information than just technical signaling protocol and parameter information. GR-145-CORE addresses interconnection from the perspective of how the LEC is or is not capable of providing information. For example, GR-145-CORE describes how the LEC provides Feature Group A, B, C, and D support. Essentially, GR-145-CORE is an implementation document for LEC interconnection where the LEC is at the center of the interconnection view. GR-145-CORE also promotes Bellcore document products. TIA's IS-93 addresses the interconnection not only from an implementation perspective, but also from the perspective of a wireless carrier "wish list" where the wireless carrier is at the center of the interconnection view; the wireless carrier is assumed to be a network to which other carriers wish to interconnect.

Both documents are undergoing rapid changes in order to maintain market relevance. The following information may be out of date by the time of this book's publication.

Given its long history, Bellcore's GR-145-CORE is still the predominantly quoted industry specification. The documents are officially titled:

- EIA/TIA IS-93 Revision A, September 1998, "Cellular Radio Telecommunications A_i-D_i Interfaces Standard."
- GR-145-CORE Issue 1, March 1996, "Compatibility Information for Interconnection of a Wireless Services Provider and a Local Exchange Carrier Network."

INTERFACE TYPES

GR-145-CORE focuses on 10 specific categories of interconnections. IS-93 focuses on 13 specific categories of interconnection. The categories of both documents are largely the same from a protocol perspective. GR-145-CORE supports the following categories:

1. Direct *Wireless Service Provider* (WSP) connection through a LEC end office using *Multi-Frequency* (MF) signaling, called Type 1.
2. Direct WSP connection through a LEC end office using the *Integrated Services Digital Network* (ISDN) protocol, also known as Type 1 with ISDN.
3. Direct WSP connection with a LEC tandem office using MF signaling, called Type 2A.
4. Direct WSP connection with a LEC tandem office using SS7 signaling, called Type 2A with SS7.

5. Direct WSP connection with a LEC *Common Channel Signaling* (CCS) *Signaling Transfer Point* (STP), called Type S.
6. Direct WSP connection with a specific LEC end office using MF signaling, called Type 2B.
7. Direct WSP connection with a specific LEC end office using SS7 signaling, called Type 2B with SS7.
8. Direct WSP connection with a LEC tandem office arranged for 911 emergency calls, called Type 2C.
9. Direct WSP connection with a LEC tandem office arranged for LEC operator assisted calls or directory service using MF signaling, called Type 2D.
10. Direct WSP connection with a LEC tandem office arranged for LEC operator assisted calls or directory service using SS7 signaling, called Type 2D with SS7.

TIA's IS-93 supports the following categories:

1. Trunk with Line Treatment using MF signaling, called *Point of Interface* (POI) -T1.
2. General Trunk Access Signaling using MF signaling, called POI-T4.
3. General Trunk Access Signaling using SS7 signaling, called POI-T5 and POI-S5.
4. Direct Trunk Access Signaling using MF signaling, called POI-T6.
5. Direct Trunk Access Signaling using SS7 signaling, called POI-T7 and POI-S7.
6. Operator Services Access Signaling using MF signaling, called POI-T10.
7. Operator Services Access Signaling using SS7 signaling, called POI-T11 and POI-S11.
8. Call Management Features Signaling using MF signaling, called POI-T12.

9. Call Management Features Signaling using SS7 signaling, called POI-T13 and POI-S13.
10. Basic Signaling Transport using SS7 signaling, called POI-S14.
11. Global Title Signaling Transport using SS7 signaling, called POI-S15.
12. Cellular Nationwide Roaming Signaling using SS7 signaling, called POI-S16.
13. TCAP Applications using SS7 signaling, called POI-S17.

Bibliography

Azmak, Okan, (1996-2000), telecommunications expert. Senior network architect for Flash Networks.

Bates, Regis J. and Gregory, Donald (1998), "Voice and Data Communications Handbook,, McGraw-Hill.

Berson, Alex; Smith, Stephen; and Thearling, Kurt, (2000), "Building Data Mining Applications for CRM," McGraw-Hill.

Chu, Lawrence (1984-1999), currently President of Mediacom Ventures Consulting. Consolidated work and teachings of L. Chu, New York Telephone, NYNEX, and Bell Atlantic-engineering and regulatory.

Eifinger, Charles P. (1984-2000), consolidated teachings of Charles Eifinger, telecomm engineering consultant.

Feit, Sidnie Dr. (1999), "TCP/IP," McGraw-Hill.

Fix, Michael S. (1996-1999), United States Air Force retired, former Director of engineering for Excel, currently Director of Technology for Lucent.

Goralski, Walter J. and Kolon, Matthew C. (2000), "IP Telephony," McGraw-Hill.

Jeffrey, Stu (1998-2000), telecommunications expert

Kotler, Philip (1997), "Marketing Management," Prentice-Hall.

McGrath Hadwen, Eileen (1998-2000), consolidated work. President of McGrath Hadwen Associates; Boulder, CO.

Notes on the Network 1980, AT&T.

"Official Wireless Application Protocol," by Wireless Application Protocol Forum, Ltd.; Wiley.

O'Neill, James (1999-2000), former vice president for ITDS, currently international billing consultant.

Proffitt, James (2000), former vice president of engineering for AirTouch Cellular and Verizon Wireless. Consolidated work of J. Proffitt.

Russell, Alicia (1996-2000), formerly of SBC Corporation, currently Director for Technology for Mobileum. Consolidated work of A. Russell.

Spragg, Mark (2000), Price Waterhouse Coopers.

St. Laurent, Simon and DeLong, B.K. (2000), "XHTML, Moving Towards XML," M&T Publishing.

Standing, Craig (2000), "Internet Commerce Development," Artech House.

Taylor, Carlyn (2000), Price Waterhouse Coopers.

Telecommunication Transmission Engineering (1977), AT&T.

Wilder, Floyd, Guide to the TCP/IP Protocol Suite, (1993), ARTECH House.

Work of the World Wide Web Consortium, (1998-2000).

Young, Harry (1992-1999), consolidated work of Harry Young, interconnection consultant.

INDEX

B

F

G

H

I

K–L

M

Q–R

S

X

About the Author

P.J. Louis has nearly a quarter of a century's worth of experience in the telecom business. Mr. Louis is currently Vice President of Carrier Marketing & Product Management with TruePosition, Inc., a leading provider of wireless location services. Mr. Louis had also served as chief of staff for engineering in NYNEX. He has held a number of leadership positions within Bell Communications Research and NextWave Wireless. Mr. Louis is a former officer of the *Institute of Electrical and Electronics Engineers* (IEEE) Communications Society—New York Section. Mr. Louis is a registered engineer in the State of New York. Mr. Louis is also the author of "Telecommunications Internetworking."